Advances in Anatomy Embryology and Cell Biology

Vol. 133

Springer
Berlin
Heidelberg
New York
Barcelona
Budapest
Hong Kong
London
Milan
Paris
Santa Clara
Singapore
Tokyo

I. Darian-Smith M.P. Galea
C. Darian-Smith M. Sugitani
A. Tan K. Burman

The Anatomy of Manual Dexterity

The New Connectivity of the Primate Sensorimotor Thalamus and Cerebral Cortex

With 41 Figures, 18 in Colour, and 5 Tables

Springer

Dr. Ian Darian-Smith
Department of Anatomy and Cell Biology
Howard Florey Institute of Experimental Physiology
and Medicine
University of Melbourne
3052 Parkville, Victoria
Australia

Dr. Mary P. Galea
Dr. Corinna Darian-Smith
Dr. Michio Sugitani
Dr. Andrew Tan
Dr. Kathleen Burman

Brain Research Laboratory
Department of Anatomy and Cell Biology
University of Melbourne
3052 Parkville, Victoria
Australia

ISBN 3-540-61111-8 Springer-Verlag Berlin Heidelberg New York

Library of Congress Cataloging-in-Publication Data. The anatomy of manual dexterity; the new connectivity of the primate sensorimotor thalamus and cerebral cortex/Ian Darian-Smith ... [et al.]. p. cm.—(Advances in anatomy, embryology, and cell biology; v. 133) Includes bibliographical references and index. ISBN 3-540-61111-8 (softcover) 1. Thalamus—Physiology. 2. Sensorimotor cortex—Physiology. 3. Neural circuitry. 4. Motor ability. 5. Macaques—Physiology. I. Darian-Smith, I. (Ian) II. Series. QL801.E67 vol. 133 [QP383.5] 574.4 s—dc20 [599.8′04188] 96-17837

Printed in Germany

Cover design: Design & Production GmbH, Heidelberg

Typesetting: Best-set Typesetter Ltd., Hong Kong

SPIN: 10522876 27/3136/SPS – 5 4 3 2 1 0 – Printed on acid-free paper

Acknowledgements

The University of Melbourne through the Faculty of Medicine and the Department of Anatomy and Cell Biology, the National Health and Medical Research Council, Australia, the Howard Florey Institute, and the Ian Potter Foundation have all generously supported the work of the Brain Research Laboratory over a number of years. All members of the Laboratory have contributed greatly to the work on which this review is focused, ranging from the care and training of the monkeys used, the histochemistry, and the computer-based analysis of data and the construction of graphics. M.P. Galea prepared the data for most of the graphics relating to corticospinal neuron populations, Corinna Darian-Smith prepared most of the graphics concerned with the topography of the macaque's thalamus, and A. Tan was responsible for the illustrations of the morphology of injected thalamocortical and corticothalamic neurons. Finally, M. Sugitani was primarily responsible for setting up the facilities for injecting prelabeled neurons by iontophoresis, as reported earlier by Buhl and Lubke (1989).

Contents

1 Introduction

1.1 Purpose and Plan of This Review

This review is focused on the topography and connections of some of the neuron populations that determine the manual dexterity of the macaque monkey. The populations selected for examination are the following:

1. The corticospinal neuron populations
2. The thalamocortical and corticothalamic neuron populations associated with the sensorimotor cortex
3. The ipsilateral cortical connections of the sensorimotor cortex

These neuron populations have been chosen because of their obvious relevance to the directed, intelligent use of the hands, but also because of their anatomical and functional interdependence. Corticospinal neuron populations transmit a complex, orchestrated output from a number of different regions of cerebral cortex to the neuron populations in every segment of the spinal cord, and this output includes the command information defining the intended manual action. The thalamocortical complex is especially concerned with the transmission and modulation or filtering of (a) visual, tactile, proprioceptive, vestibular, and auditory information to the cerebral cortex and (b) information from the cerebellum, basal ganglia, limbic system, and brain stem which is relevant to sensorimotor behavior. Finally, the extensive ipsilateral cortical connections constitute a major part of the supraspinal circuitry which coordinates the contributions of all the cortical neuron populations contributing to intelligent sensorimotor behavior and, in particular, transmits the cross talk between those cortical neuron populations which shape and control the dextrous handling of objects within reach. In order to contain the review, interhemispheric connections are not considered in detail. For the same reason cerebellar connections with the cerebral cortex and spinal cord are reviewed within the framework of the three chapters outlined above, and not separately.

In the last decade, each of these neuron populations has been studied in some detail using modern axon tracer procedures, single neuron recording in the active alert monkey, positron emission tomography (PET), and the sensorimotor behavioral changes resulting from their selective destruction. Some real advances have been made, both in the description of these neuron populations and also in our understanding of their operation. The special importance of these recent studies is that they have initiated a serious reconsideration of the structural and functional organization of the sensorimotor neuron populations mediating voluntary action. Our goal has been to review these experimental findings and to critically examine some of the new ideas triggered by these observations.

1.2 Primate Manual Dexterity

1.2.1 Using the Hand

The intelligent use of the hand as a sensorimotor organ to explore the visible world and to locate, identify, and appropriately use objects within reach has been an important factor contributing to the biological success of the primates among mammals (Napier 1980). Of course, not all primates are equally dextrous. Man has a most adaptable hand, capable of opposing the pad of the thumb with those of each of the other digits and of moving and applying forces with each finger independently of its neighbors, so that the form, texture, and consistency of handled objects may be identified. There is also some "sharing of labor," expressed as handedness, so that some sophisticated manual tasks, such as writing, are better executed by a particular hand. Humans also use and make tools, which extend their capacity for manipulating objects well beyond that achieved by the apes. Using objects as tools is not uniquely human, nor even uniquely mammalian, but toolmaking, in which an object is modified and used regularly in order to extend and execute a particular manual task, is a primate feature and, as Benjamin Franklin noted (Napier 1980), certainly best developed in man. The macaque, which is the focus of much of this review, is, unlike the apes, still a quadruped and does not make tools. Although displaying a preference for using one or other hand, the macaque lacks the handedness of the human subject and also the capacity to truly oppose the terminal pads of the thumb and other digits. Not all fingers in the macaque can be moved completely independently of their neighbors. Nonetheless, the monkey uses a whole repertoire of digital movements to grasp objects, explore their surfaces, and identify their shape and surface texture in a way that approximates that of humans. Macaques may even use objects as simple tools (Richard 1985). For this reason, Kavanagh (1983) referred to the cercopithecines, which include the macaques, as "intelligent manipulators."

Figure 1.1 illustrates some features of a simple manual task that is executed hundreds of times daily throughout the life of the monkey, namely, reaching out and retrieving a small solid object within the subject's view (see Fig. 1.4; note that the latter illustration is of manual retrieval in a monkey following a spinal cord lesion and differs in detail from behavior in the normal macaque; see below). In order to retrieve the horizontally oriented cylindrical object (5 mm in diameter) with the index finger and thumb, these digits must be moved in the vertical plane. In the normal macaque this vertical alignment usually occurs soon after beginning to reach for the object. The black lines (Fig. 1.1) in this vertical plane represent the paths of the pads of the thumb and index finger, and together they define the *opposition space* (Jeannerod 1994a,b; Jeannerod et al. 1994) in the vertical plane, which reflects the preshaping of the thumb and index fingers, and the change in wrist angle, prior to grasping the target object. This preshaping of the hand, thumb, and index fingers is determined partly by the feedback of visual

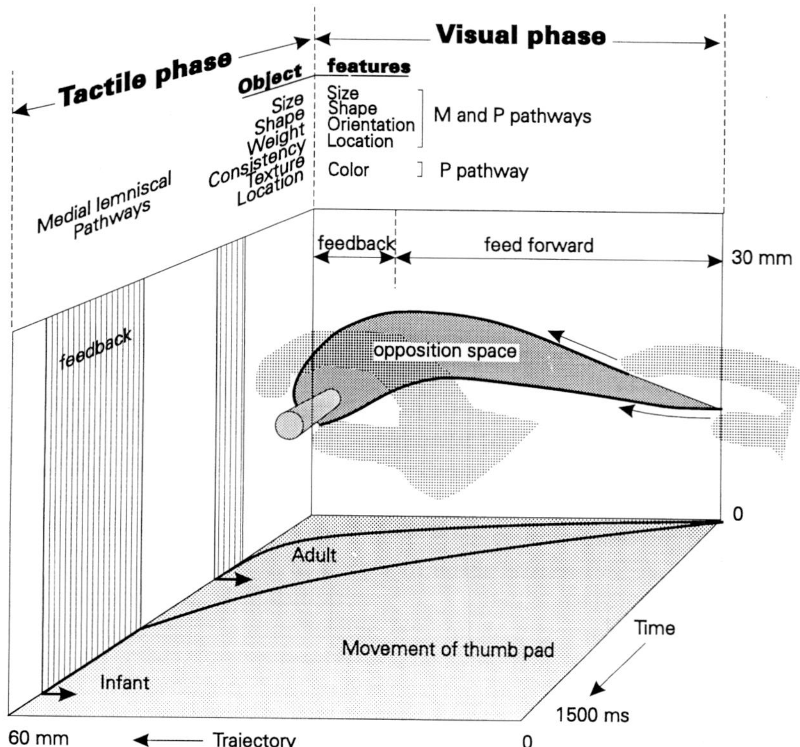

Fig. 1.1. Reach and grasp behavior in macaque monkey. Trajectories of index finger and thumb pads outline the *shaded* opposition space in the vertical plane and the separation of the pads prior to grasping the cylindrical target object. The visual phase ends and the tactile phase starts when the index pad contacts the target; both phases have feedforward and feedback components, which vary in importance, depending on the speed and complexity of the task. Information reaching the brain about some features of the target (e.g., location in external world, color of object) is purely visual; some object features are purely tangible (such as physical consistency and weight), whereas most features, such as size, shape, surface texture, and orientation of the object, are assessed from both visual and tactual information. The timing of the thumb trajectory for grasping the same object in both the 3-month-old and adult macaque is plotted on the floor of the figure

information concerning the location, size, shape, and orientation of the target object and the spatial relations of the hand and target object. The hand does not have to be visible. The black lines plotted on the shaded horizontal plane relate the position of the pad of the thumb to the time after the onset of the reaching movement; this is shown for both the mature and infant macaque. Two phases of the reaching and grasping movements can be somewhat arbitrarily differentiated (Jeannerod et al. 1994) by the fact that the first is visually guided ("visual" phase), and the second starts when one or more digits contact the target ("tactile" phase). The monkey "leads" with the index finger, which traverses a curved trajectory to contact the target on its distal surface. The extended thumb traverses a similar path until it contacts the target. The visual phase can be divided into an initial ballistic ("feedforward") component and a later slow, "accurate" phase, which is mainly guided by the feedback of visual information. The tactile phase is largely regulated by the feedback of tactile and kinesthetic information about the size, shape, weight, and consistency of the target object. If, as in Fig. 1.1, the target object is held by a spring clip, then this tactile/proprioceptive information is also used to adjust the opposition force applied to retrieve the object.

In daily activities, the manual retrieval described above is not usually performed in isolation, but within the context of elaborate purposeful behavior, with, for example, the object being placed in the mouth, or discarded, or used as a tool. Nonetheless, even this simple task involves sequences of movements that occur coincidentally, which must be coordinated, and which may well be regulated by separate control systems. Several investigators (Jeannerod 1994a,b; Stelmach et al. 1994; Arbib and Hoff 1994; Arbib et al. 1985; Iberall et al. 1986) have developed models of the coordinated processes that underly manual retrieval and, in turn, related these to neuronal processing in the sensorimotor cortex. The different models that have been proposed have some common features, in that they each propose a number of control systems operating in parallel, with varying bases and degrees of correlation and interaction. There are too many independent behavioral events occurring simultaneously for them to be described in terms of a realistic serial model of processing.

For the macaque to execute the reach-and-grasp movement shown in Fig. 1.1, it must continuously define in neural terms the surrounding three-dimensional world; this assessment is achieved by combining information derived from multiple sources concerning the different physical and biological variables. Visual information helps in defining *what* the target is and *where* it is located relative to both the external world and to the hand and other body parts (visual phase). The complementing proprioceptive, vestibular, and tactile information signals where the arm, hand, and fingers are relative to the rest of the body (McCloskey 1994), and once the fingers touch the object it specifies the details of its mass and form, surface texture, and compliance (tactile phase). Within the space defined by the sum of this information, the monkey must be able to move the hand and fingers quickly and precisely to execute the desired task. Moreover, even with the simple retrieval task illustrated in Fig. 1.1, other behavioral processes have preceded it and also overlap with its execution. The monkey has selectively attended to the target object, recognized it to be food or something else of interest, decided to grasp and use it, and "planned" the necessary motor task, taking into account the position of the object relative to the trunk, head, and hand. As has become clear in PET and functional magnetic resonance imaging (MRI) studies in humans, each of these preliminary or coincident behavioral events may be reflected in the topographic distribution of the regions of cerebral cortex that are active during the manipulation (Seitz et al. 1991; Deiber et al. 1991; Grafton et al. 1992; Shibasaki et al. 1993; Wessel et al. 1994; Remy et al. 1994). From these considerations it is apparent that the slow and uncertain reaching and handling of an object by an infant monkey, as shown in Fig. 1.1, may result not only from incomplete motor signals being relayed from the cerebral cortex to the spinal cord, but also from an incomplete visual and somesthetic image of the continuously changing world in which the monkey operates, from slow recognition of the target object, or from a slowness in formulating the purpose and plan of the manual task. In infant monkeys, the feedback of visual and tactile information about this world, full of novel and unrecognized objects, probably subserves a somewhat different role than it does in the mature animal. The latter can recall from previous experience information concerning a familiar external world or target object; in such circumstances, it may depend less on direct sensory feedback than the infant does.

1.2.2 Cerebral Cortex and Dexterity

It has been known for a century that the execution of any directed, intelligent motor act depends on the functional integrity of particular areas of the cerebral cortex. During that century there has been a repeated revision of ideas concerning which neuron populations of the neocortex (the "sensorimotor cortex") contribute to the execution of voluntary behavior, how these different cortical populations operate together, how they interact with subcortical neuron populations in the thalamus, basal ganglia, brain stem, cerebellum, and spinal cord, and as a result generate the patterns of motoneuron activity that elicit the intended movements. The trend has been to identify successively more extensive and more widely distributed cortical neuron populations which contribute to sensorimotor behavior and to increasingly appreciate the complexity of even the more experimentally accessible neuronal processes mediating voluntary action. In the last 15 years, with the rapid development and application of modern axon tracers to the analysis of the monkey's sensorimotor pathways (Kievit and Kuypers 1977; Jones and Friedman 1982; Asanuma et al. 1983a; Jones 1985; Toyoshima and Sakai 1982; Hutchins et al. 1988; Goldman-Rakic 1988a,b; Dum and Strick 1991, 1992; Darian-Smith et al. 1990a,b, 1993; Holsapple et al. 1991; Galea and Darian-Smith 1994, 1995), the direct correlation of single cortical neuron responses with the experimental monkey's own manual performance (Evarts 1966; Evarts et al. 1984; Mountcastle et al. 1975, 1978; Kalaska and Crammond 1992; Lawrence 1994; Schwartz et al. 1988; Darian-Smith et al. 1984, 1985; Georgopoulos 1986; Georgopoulos et al. 1984; Andersen 1989; Andersen et al. 1993a,b), and the recent development of PET and functional MRI (Grafton et al. 1992; Friston et al. 1993; Colebatch et al. 1991; Raichle et al. 1994; Burton et al. 1993), there has been a substantial acceleration in the identification of the connections of the sensorimotor thalamus and cerebral cortex in the primate. Investigators have become increasingly aware of the difficulty in defining a functional concept, the sensorimotor cortex, in anatomical terms. Clearly, the extent of the active sensorimotor cortex depends on the complexity of the sensorimotor behavior it mediates; in other words, the sensorimotor cortex is not a static, anatomical entity, but must be defined in relation to the sensorimotor task being examined. In this review we use the term "sensorimotor" cortex (and "Sensorimotor" thalamus) in this functional sense.

After J.H. Jackson's (1875, 1897) initial clinical correlation of focal "motor" and "sensory" epilepsy with lesions in the pre- and postcentral gyri of the human cortex, respectively, Ferrier (1876) and others (Leyton and Sherrington 1917; Foerster and Gagel 1932; Penfield and Boldrey 1937; Woolsey et al. 1952; Asanuma and Rosen 1972; Sessle and Wiesendanger 1982) mapped out the electrically excitable "motor" cortex of the macaque monkey and human. This included an area in the lateral precentral gyrus, from which contralateral hand and finger movements could be elicited by electrical stimulation, but also other similar but separate areas in the medial frontal cortex (supplementary motor area, SMA), the postarcuate cortex, the insular cortex, and recently the anterior cingulate cortex (Matsuzaka et al. 1992; Tanji 1994). During this same period, *direct* corticospinal projections (Leyton and Sherrington 1917; Phillips and Porter 1977; Porter and Lemon 1993, for reviews; Pandya et al. 1981; Porrino and Goldman-Rakic 1982; Hutchins et al. 1988; Galea and Darian-Smith 1994, 1995) and indirect projections from the cortex to spinal cord, with relays in the red nucleus, and brain stem reticular formation (Kuypers and Lawrence 1967; Kuypers 1981; Humphrey et al. 1984; Humphrey and Tanji 1991) were identified. Figures 2.2 and 2.11 illustrate in much simplified diagrams the direct and some of the indirect corticospinal projections in the primate, which will be considered further in Chap. 2.

By the beginning of the present century, the major *afferent* projections to the primate sensorimotor cortex had also been identified, but little was known of their functional architecture or physiology. The thalamus was recognized to transmit virtually all of the information received by the neocortex from sense organs and other regions of the brain. Subsequent important steps in understanding thalamocortical organization in the primate included the following:

1. The cytoarchitectonic parcellation of the thalamus (LeGros Clark 1932; Walker 1938; Olszewski 1952; Jones 1985) and of the frontal and parietal cortex (Betz 1874; Campbell 1905; Brodmann 1909; Vogt and Vogt 1919; Von Bonin and Bailey 1947)
2. The effects of focal lesions of the frontal and parietal cortex and their projections on movement control and somatic sensibility (Head and Holmes 1911; Head 1918; Ruch et al. 1938; Kennard 1938, 1940; Lawrence and Kuypers 1968a,b)
3. The analysis of single neuron responses in the macaque's somatosensory thalamus (Mountcastle and Henneman 1952; Poggio and Mountcastle 1960; Mountcastle and Poggio 1963; Mountcastle 1984) and in pericentral cortex (areas 4, 3a, 3b, 1, 2, 5, and 7) of the alert, active macaque (Evarts 1966; Mountcastle et al. 1975; Phillips and Porter 1977; Porter and Lemon 1993; Kaas et al. 1979; Nelson et al. 1980; Hyvarinen 1982; Jones and Friedman 1982; Georgopoulos et al. 1984, 1993; Darian-Smith et al. 1984, 1985)
4. Early studies using axonally transported tracers for identifying projections in the macaque (Jones and Powell 1970b; Gatter and Powell 1978; Kievit and Kuypers 1977)

Figures 1.2 and 1.3 summarize the topography of the major thalamic nuclei in the macaque and their cortical projections (Olszewski 1952; Jones 1985). Olszewski's coronal cytoarchitectonic maps were used to construct the three-dimensional thalamus; the graphics were edited to correct some inconsistencies in the original maps (see Darian-Smith et al. 1990a) and to incorporate Jones' proposed fusion of Olszewski's nuclei VPLo, VLc, VLps, and X (see Table 3.1) to constitute a single nucleus, VLp, specified by its cerebellar input. Olszewski's two-dimensional maps (1952) of the brain of the rhesus monkey (*Macaca*

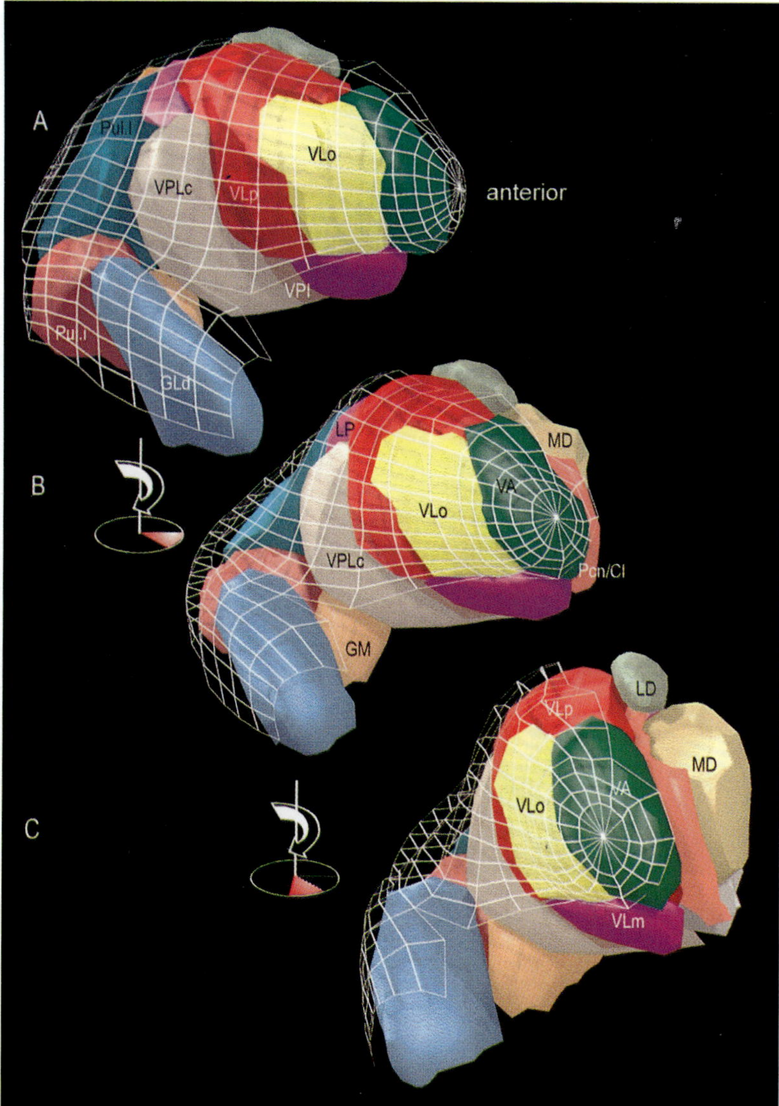

Fig. 1.2A–C. Three-dimensional reconstruction of nuclei of right dorsal thalamus of *Macaca mulatta*, constructed from the 23 coronal sections of Olszewski's (1952) cytoarchitectonic atlas. The thalamus has been rotated in two steps from a lateral view (**A**) to an intermediate (**B**) and then a frontal view (**C**). Olszewski's nuclear classifcation has been used, except that Jones' fusion of VLc, VPLo, VLps, and X (see Table 3.1) into the one nucleus VLp has been adopted, and other minor changes have been made (Darian-Smith et al. 1993). Thalamic maps of the cytoarchitectonic boundaries of nuclei commonly endow them with a sharpness which they lack in the original Nissl sections; this is also true for this three-dimensional reconstruction. However, in the three-dimensional map the visualized nuclear boundaries represent data from a number of serial sections, rather than from single sections. Abbreviations are mainly those used by Olszewski (1952): *Pul.l*, lateral pulvinar; *VPLc*, caudal lateral ventral posterior nucleus; *VLp*, ventral lateral posterior nucleus; *VLo*, oral ventral lateral nucleus; *VPI*, inferior ventral posterior nucleus; *Pul.i*, inferior pulvinar; *GLd*, lateral geniculate nucleus; *LP*, posterior lateral nucleus; *MD*, medial dorsal nucleus; *VA*, anterior ventral nucleus; *GM*, medial geniculate nucleus; *Pcn/Cl*, paracentral/central lateral nucleus; *LD*, dorsal lateral nucleus; *VLm*, medial ventral lateral nucleus. The color coding of thalamic nuclei assists in identifying the cortical areas receiving input from them, as mapped in Fig. 1.3. The subthalamic connections of some of these thalamic nuclei are illustrated in Figs. 3.2, 3.3, and 3.4

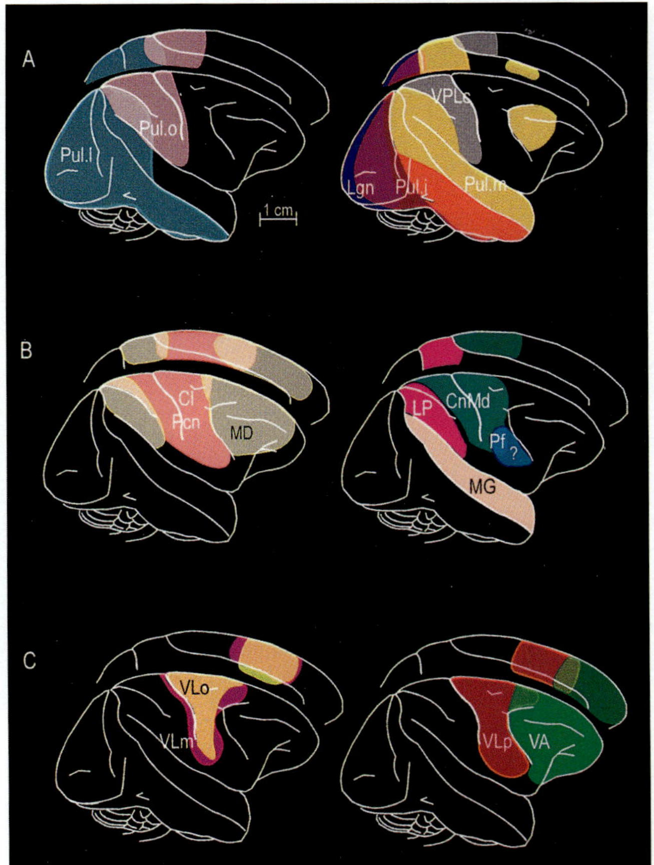

Fig. 1.3A–C. Thalamic nuclear projections to the cerebral cortex in the macaque. The color coding corresponds to that used in Fig. 1.2 for thalamic nuclei. Thalamic nuclei are labeled. *Pul.l*, lateral pulvinar; *Pul.o*, oral pulvinar; *VPLc*, caudal lateral ventral posterior nucleus; *Lgn*, lateral geniculate nucleus; *Pul.i*, inferior pulvinar; *Pul.m*, medial pulvinar; *Cl*, central lateral nucleus; *Pcn*, paracentral nucleus; *MD*, medial dorsal nucleus; *LP*, posterior lateral nucleus; *CnMd*, medial central nucleus; *Pf*, parafascicular nucleus; *MG*, medial geniculate nucleus; *VLo*, oral lateral ventral posterior nucleus; *VLm*, medial ventral lateral nucleus; *VLp*, ventral lateral posterior nucleus; *VA*, anterior ventral nucleus. Note the substantial overlap of thalamic projections; each localized region of cortex is the target of at least two nuclei in the dorsal thalamus. The medial surface of the right hemisphere is represented by its mirror image; all the projections shown are ipsilateral. See Fig. 3.1 for the different cytoarchitectonic cortical areas in the macaque brain

mulatta) have been used to construct this figure so that the reader can compare the present three-dimensional reconstruction with the "raw" data on which it is based. However, maps were constructed in our laboratory of the thalamus of two other species, *M. nemestrina* and *M. fascicularis*, and, except for their different volumes, found to be similar. The cortical fields of the projection from each thalamic nucleus are based on a composite of data from a large number of papers (for review of literature up to 1985, see Jones 1985; for studies of thalamocortical projections in the macaque since 1985, see Chap. 3 of this review). The thalamic and cortical maps of Figs. 1.2 and 1.3 highlight two aspects of their connections. First, neuron populations in the thalamus that contribute to the monkey's manual dexterity are widely distributed in a large fraction of all its cytoarchitectonically differentiable nuclei. Second, as with the thalamus, there are cortical neuron populations which contribute to visually guided voluntary arm/hand/finger movements distributed in many of the cytoarchitectonically distinct areas within the frontal, parietal, temporal, and even occipital cortex. Figure 1.3 illustrates the extent of thalamic connections in the macaque, but not its somatotopic organization; this is considered in Chap. 3.

Two concepts, which now need considerable revision, were clearly expressed in the early models of the neuronal processes mediating sensorimotor behavior. *Sequential behavior* – reaching, handling, using an object, or typing the letters of a word – was seen to result from *serial neuronal processing*, with the different sensory systems transmitting information about the external and intrinsic space first to the specific sensory cortical areas and then, through cortical connections, to particular areas within the parietal or temporal "association" cortex. After due transformation, this information was then thought to be relayed to the prefrontal cortex by specific corticocortical connections, and from there fed back into the primary motor cortex. The cognitive component of the motor action, its planning and initiation, was considered to develop mainly in this prefrontal cortex and to possibly be further transformed in SMA. The essential executive commands for activating the limb and other muscles to produce desired patterns of movement were seen to be formulated in the motor cortex. Finally, these "motor" signals were transmitted to the spinal motoneuron populations through direct and indirect corticospinal pathways.

The second important idea embedded in these early models of sequential sensorimotor neuronal processing, which must now be questioned, was the *local representation of information* within topographically identified neuron populations. In particular, the executive commands for the spinal motoneuron populations would be transmitted from a particular region of the precentral primary motor cortex. This concept of localization of function within the cerebral cortex has a long history, being given a jump-start by Ferrier's mapping (1876) of the somatotopically organized, electrically excitable cerebral cortex in the macaque. The correlation of particular focal parietal and frontal cortical lesions by Broca (1861) and then Wernicke (1874) in the newly identified "dominant" cerebral hemisphere in humans with particular types of aphasia greatly reinforced this idea of the neuron populations of a particular area of cortex subserving a particular aspect of cognitive or sensorimotor behavior. Subsequent clinical studies of the behavioral deficits resulting from focal cortical lesions of the somatosensory postcentral cortex (see above), of the agnosias resulting from posterior parietal lesions ("astereognosis"), and of more complex cognitive dysfunction following localized lesions of the prefrontal and temporal cortex gave credence and respectability to this idea of the localization of function in specific areas of cerebral cortex. The latter concept had been inauspiciously launched earlier in the nineteenth century in the writings of Gall and Spurzheim (1809) on phrenology (see Zola-Morgan 1995).

1.2.3 Topography of Sensorimotor Pathways that Mediate Handling

Even in the early writings on the the neuronal processes mediating voluntary action, there was some questioning of the appropriateness of a model based solely on the *serial* processing of information and on its *local representation* within topographically specified neuron populations. Embedded in Hughlings Jackson's somewhat convoluted analysis of levels of operation of the brain (e.g., Croonian Lectures, Jackson 1875, 1897) was his appreciation that apparently serial or sequential behavior does *not* necessarily imply sequential neuronal processing of information. Perhaps the most unsettling finding that Hughlings Jackson and subsequent investigators found difficult to explain in terms of a serial model of informational processing is the remarkable resilience of voluntary sensorimotor behavior when these known pathways are locally and permanently interrupted by disease or experimental lesion. A rapid recovery of voluntary action is often seen in capsular stroke patients even when the permanent destruction of many pyramidal fibers in the internal capsule can be demonstrated by PET (Donnan et al. 1991; Weiller et al. 1992, 1993; Weder et al. 1994). Likewise, substantial behavioral recovery after an interval of weeks or months is the rule in macaques following carefully defined permanent experimental focal lesions of the pyramids or the spinal cord (Lawrence and Kuypers 1968a,b; Galea and Darian-Smith 1995; see below). Similarly, with permanent focal lesions of the postcentral somatosensory cortex in human subjects (Holmes and May 1909; Head and Holmes 1911; Holmes 1926; Semmes et al. 1963), complete loss of

Fig. 1.4. Reach and grasp in a macaque monkey (*Macaca nemestrina*) following cervical spinal "hemisection," which was done 14 months after birth, at C4 on the left side. After an initial left hemiplegia, the monkey recovered rapidly and within 2 months was agile in a large gang cage. Careful observation was necessary to detect any impairment of the monkey's movements about the cage. Four months after the hemisection, a stereotyped reach-and-grasp task requiring the opposition of the thumb and index finger to remove a 5-mm target object against a resistance of approximately 1 N was performed with the left hand, but was slower than when performed with the right hand. These movements are shown in the photographs (*left*, left hand; *right*, right hand). Finger/thumb opposition was also weaker on the left side, so that increasing the resistance to moving the target to more than 2 N resulted in a marked slowing of the performance with the left hand, but not the right. There was some independent digital movement in the left hand, with precise movements of the index finger being used to hook it around the target object, in order to draw it out of the spring clip against the flexed thumb. However, the opposition space, defined by the trajectories of the left index and thumb (see Fig. 1.1), was less well matched ("preshaped") to the dimensions of the target object than that of the right hand. The section of the spinal cord at the level of the original hemisection (luxol fast blue stain, *bottom right*) shows marked left-sided atrophy and that the hemisection missed only part of the fasciculus gracilis and of the ventromedial column (left side of spinal cord shown on right side of section). No fibers descending in the left cord below the lesion were detected with anterograde labeling. However, at every segment caudal to the lesion, both anterograde and retrograde labeling showed that there were fibers projecting from the sensorimotor cortex of both hemispheres to the left spinal cord, as shown in the diagram in Fig. 1.5; all these fibers traversed the cervical spinal cord on the right side and crossed the midline caudal to the level of the lesion. In other experiments in which the cervical spinal cord was hemisected, the retrograde labeling of descending corticospinal fibers showed that these fibers were only about 10% of the total corticospinal projection which descends in the dorsolateral column to the ▶

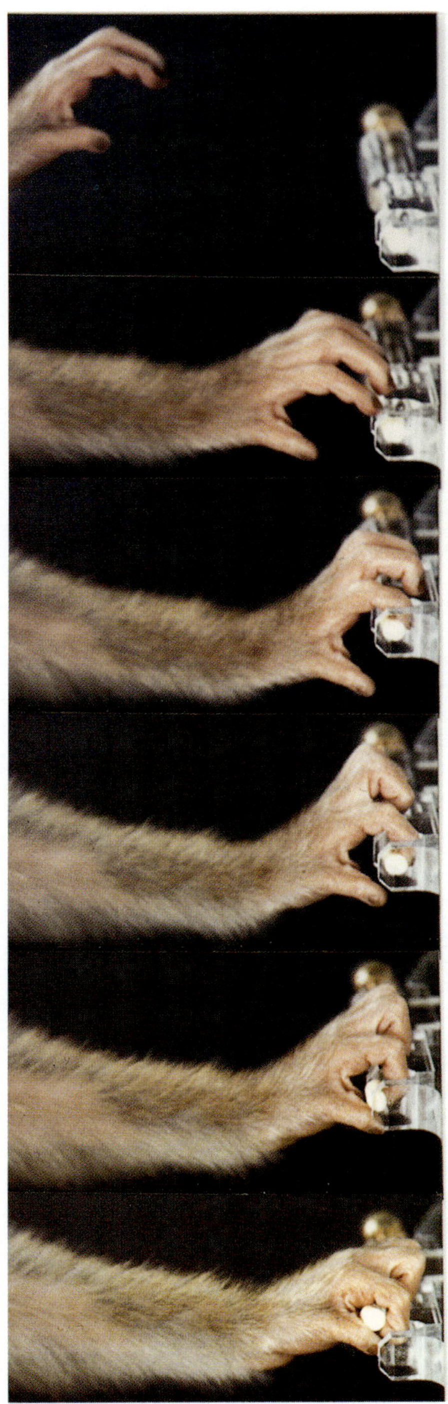

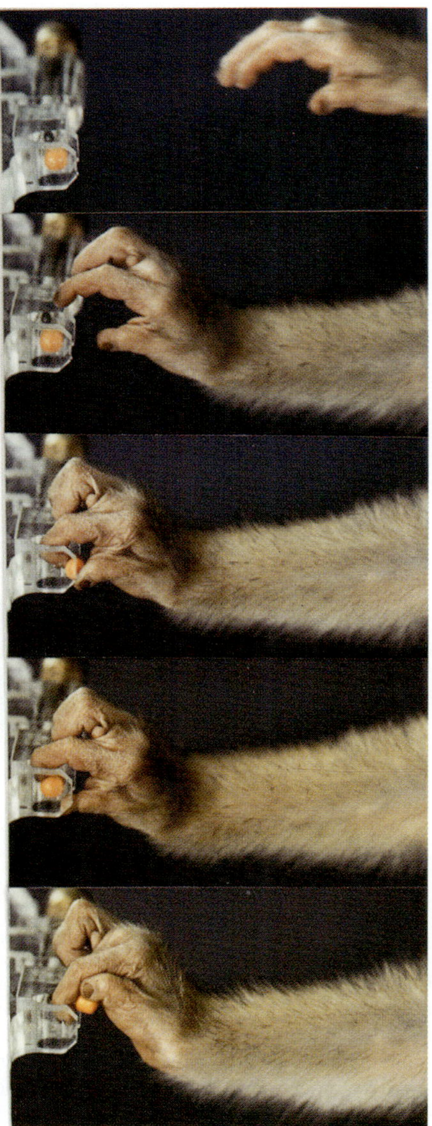

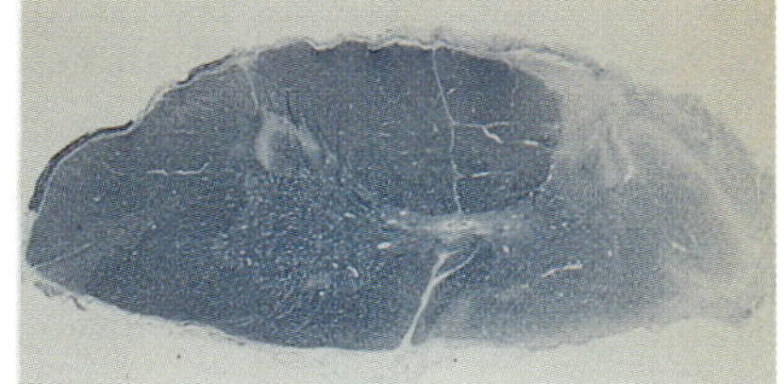

contralateral spinal cord in the normal macaque (Galea and Darian-Smith 1995). Thus permanent slight loss of complex finger movements and applied forces does occur with a 90% reduction in the direct corticospinal projection, but the recovery of manual dexterity that occurs is remarkable

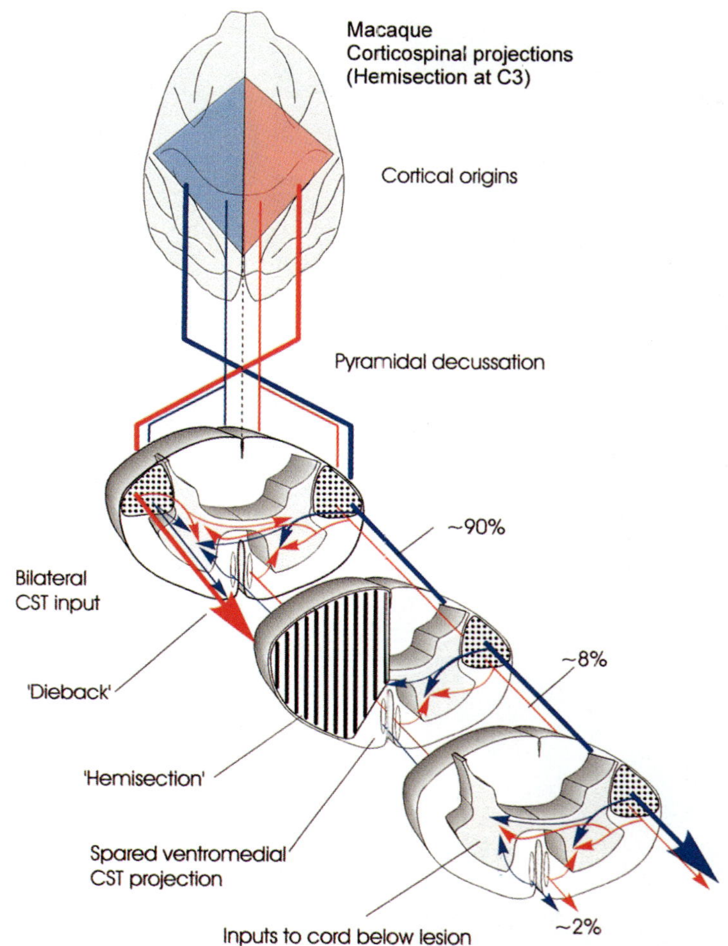

Fig. 1.5. Corticospinal projections in adolescent macaques following a spinal "hemisection" at about C4. This diagram is based on a large series of macaques (Galea and Darian-Smith 1995), including the monkey whose post-section manual dexterity is illustrated in Fig. 1.4. In all monkeys the hemisection completely interrupted the dorsolateral column; in most monkeys small segments of the fasciculus gracilis and of the ventromedial column were spared by the lesion, but in some either the dorsal column or the ventromedial column was completely interrupted. No anatomical reconstruction of the severed pathways could be detected using anterograde and retrograde axonal tracers in any of our experiments. This was so in monkeys in which the spinal cord was hemisected at varying ages (5 days to 2 years), with the recovery phase examined at time intervals after spinal section ranging from 2 weeks to 35 months. *CST*, cortiocospinal tract

cutaneous and proprioceptive sensibility is not observed, and in fact any demonstrable loss of somatic sensibility is usually slight.

The above comments do not imply that complete recovery of sensorimotor function occurs following permanent focal damage of major sensorimotor pathways. Some residual permanent loss of manual dexterity can usually be demonstrated if searched for, and it is probably always present if the monkey is examined sufficiently carefully. What is striking, however, is that even with large, well-defined lesions of these various pathways, the functional recovery of voluntary action in the patient or experimental monkey is often substantial and rapid. This is so even though there may be little or no evidence of actual replacement of the damaged projections as much as 2–3 years after the injury.

Figures 1.4 and 1.5 illustrate the level of recovery of hand function in an adolescent macaque that occurred in the 4 months following a near-complete hemisection on the *left* side of the cervical spinal cord (at C3) which was done 14 months after birth (Galea and Darian-Smith 1995). The only fibers surviving the hemisection which link the neocortex and left cervical spinal cord caudal to the lesion were the afferent spinothalamic projection, the small complement of corticospinal and bulbospinal fibers which descend in the right side of the spinal cord before crossing the midline below the hemisection, and a few fibers in the intact left ventromedial spinal column. The surviving descending fibers, labeled by anterograde and retrograde axon tracers (corticospinal and rubrospinal), amounted to less than 10% of their normal complement (Fig. 1.5). In spite of this extreme attrition of the direct projections linking the forebrain and spinal cord, Fig. 1.4 illustrates the remarkably effective use of the left hand and fingers in picking up a small target object (5×5 mm). This target object was held by a spring clip requiring the application of a force of approximately 1 N to release it. The trajectories of the thumb and index finger resembled normal performance in most features, including its speed, measured by the time taken to reach, grasp, and release the target, and the smoothness of execution. However, the preshaping of the digital separation to match the dimensions of the target was delayed, as is seen on comparing the trajectories of the left and right thumbs. When the spring compression force preventing release of the target was doubled (more than 2 N), the disability in the monkey's left hand was more apparent, since the force applied by the opposing digits was limited. Nonetheless, considerable independence of movement of the separate digits had returned following the extensive spinal lesion. The important finding was that within 4 months the monkey's use of the left and right hand was very similar and that further recovery of manual dexterity occurred over the following 6 months. This was the rule following cervical spinal hemisection both in the infant and mature macaque (Galea and Darian-Smith 1995). Our recent study of this recovery of hand function when the corticospinal connections have been greatly reduced highlights the following points: (a) the multiplicity of descending corticospinal pathways, of which a small but significant fraction is left intact following spinal hemisection, and (b) the very important role of spinal cord circuitry in regulating and shaping hand movements that are initiated by the forebrain. Our findings imply that substantial synaptic reorganization of the spared descending fibers is likely.

1.2.4 Parallel Distributed Processing in Sensorimotor Pathways

If the sensorimotor information needed for the manual task described above were transmitted through a succession of spatially localized neuron populations, each with unique input from relatively few sources, and each responsible for a discrete step in this processing, this "serial" transmission would be highly susceptible to focal injury at any point along the pathway Rather than this vulnerable serial circuitry, it seems more likely that some form of parallel transmission mediates sensorimotor behavior, that the transfer of relevant information is distributed across these channels, is continuous rather than staged and successive, and that these parallel transmission lines are highly interactive. Circuitry with these features is functionally much less impaired by localized damage of some of its elements than is a serial circuit. The opposing views of the transmission of information in serial and parallel neuronal pathways have passed in and out of favor for nearly two centuries (van Gelder 1984). Gowers' inference (1886) of topographically separate spinal pathways subserving touch and pain sensibility in a patient with a spinal hemisection was an early clear report of parallel (but not "distributed") transmission in the sensorimotor pathways, and this was soon followed by the recognition of indirect corticospinal projections which relay in the red nucleus and brain stem reticular formation and parallel the pyramidal tract. Nonetheless, the all-too-simple model of serial transmission in the major sensory and motor pathways was widely accepted for much of this century. Parallel transmission has recently assumed new relevance in the analysis of sensorimotor behavior, partly because of our new awareness of the complexity of the pathways mediating visually guided reaching and grasping, as well as similar sensorimotor behavior, but also because of the development of parallel computers, with their unique capacity to very rapidly execute multiple computations involving hugh amounts of information (McClelland et al. 1986; McClelland 1989; Hinton et al. 1986; Rumelhart et al. 1986; Friston et al. 1993; Nelson and Bower 1990). The apparent similarity of the simultaneous processing of many pieces of information in both the parallel computer and in neuron populations tranferring information from one part of the brain to another has had an obvious appeal. In fact, recent studies have shown that the connections between most neuron populations that operate in accord consist of several subpopulations projecting in parallel which are capable of transmitting similar, but not identical information (e.g., Selemon and Goldman-Rakic 1988; Goldman-Rakic 1988a,b; Cavada and Goldman-Rakic 1989a,b; Darian-Smith et al. 1990a, 1991b; Turman et al. 1992).

In any real nervous system, the processes that determine sensorimotor behavior include both sequential and parallel transmission of information. McClelland et al. (1986) describe these processes in terms of a *macrostructure* and a *microstructure.* Any sensorimotor behavioral task, such as reaching out and retrieving a nearby object or typing a sentence, when considered as a whole is broadly sequential. While this behavioral succession does not necessarily consist of a series of discrete steps, it does have a sequential or serial macrostructure. However, if the transition from one behavioral step in this sequence to the next is examined, even for the simplest change such as opposing the pads of the thumb and index finger at the right time or the use of different digits on the keyboard, it is evident that so many "microsteps" are required that many of them have to be completed simultaneously if the behavioral act is to be completed in the time normally taken by the human or macaque subject. In other words, the *microstructure* of the neuronal processing mediating the behavioral act must be at least partly organized in parallel in order to achieve the necessary computational efficiency.

One of the main issues examined in the following chapters of this review is the evidence of such a mix of serial and parallel distributed transmission in different neuron populations that have an important role in primate manual dexterity. The two main procedural streams for analyzing the macrostructure and microstructure of these sensorimotor pathways have been to attempt to define (a) their anatomical connectivity, which as it is currently determined experimentally at best defines the *limits* of functional interaction, and (b) their functional connectivity, estimated in terms of the physiological response characteristics of individual neurons whose location within the pathways is known and how these responses relate temporally to the activity of neurons in other sites and to the behavioral task. Correlative studies of the discharge patterns of single cortical or thalamic neurons, recorded while the monkey performs a particular manual task, provide both temporal and spatial information about the responses of a very *few* neurons. By contrast, PET and functional MRI have provided important new information about the spatial distribution of activity in large populations of (mainly) cortical neurons in the human subject during the execution of particular sensorimotor tasks, but the relative timing of these regional responses is still not known (see Friston et al. 1993 for a discussion of the different experimental correlates). The upshot is that currently we cannot relate these different measures of connectivity in any quantitative way.

In assessing the functional relevance of the transmission of related sensorimotor information in parallel, spatially separate pathways, it is important to determine (a) just how distinctive the information transferred by each such pathway really is, (b) how and to what extent the parallel transmission pathways are segregated at different topographic levels in the central nervous system, and (c) the extent of cross talk and cross-matching that can occur between them. The importance of these issues has recently been well illustrated in the analysis by Merigan and Maunsell (1993) (see also Martin 1992) of the popular model of parallel M and P pathways which, it has been proposed, transmit rather different visual information from the retina to the parietal and temporal cerebral cortex. Ungerleider and Mishkin (1982) and others (Livingstone and Hubel 1988) proposed that the so-called M pathway projecting to the parietal cortex processes visual information mainly concerned with spatial relations and the movement of objects in that space. In contrast, it was proposed that the separate P pathway relays information about objects, their form, color, and pattern not only to the parvocellular neuron populations of the lateral geniculate nucleus, but also to parts of the temporal cortex. Careful examination of experimental evidence by Merigan and Maunsell (1993) has shown that segregation of these parallel pathways from the retina to the parietal and temporal cortex is incomplete and that considerable interaction between neuron populations in the two pathways can occur. Similarly, it can be anticipated that the parallel distributed transmission in sensorimotor neuron populations does not imply complete segregation of the relevant pathways, that there is overlap in the information relayed by these parallel pathways, and that regulated interaction between them is the rule. As will be reviewed later, recent PET studies of regional cortical activity related to the use of the hand in human subjects encourage this view (Roland et al. 1980a,b; Deiber et al. 1991; Grafton et al. 1992; Shibasaki et al. 1993; Kawashima et al. 1993).

In Chap. 2 the multiple corticospinal pathways in the macaque are examined, in Chap. 3 the functional anatomy of thalamocortical connectivity is reviewed, and in the final chapter the connectivity of the macaque's sensorimotor cortex is briefy examined. Substantial advances in our understanding of each of these topics have occurred in the last decade, but the new experimental data have also defined important deficits in our knowledge. These are most apparent when attempting to interrelate the thalamocortical and corticospinal neuronal systems and to assess their contributions to the uses of the hand, as will become apparent in the following chapters.

2 Corticospinal Connections in the Primate

2.1 Introduction

Over the last 150 years, a starred cast including Hughlings Jackson, Ferrier, Betz, Sherrington, Campbell, Holmes, Brodmann, the Vogts, Foerster, Penfield, Evarts, Kuypers, and a cluster of more recent investigators have examined the parcellation of the cerebral cortex (Fig. 2.1) and corticospinal connections in man and other primates (for historical reviews, see Phillips and Porter 1977; Porter and Lemon 1993). Nonetheless, we still lack a coherent account of the organization of the constituent neuron populations and of how they contribute to the primate's manual dexterity and other voluntary movement (Fetz 1992). With the introduction and development of new retrograde and anterograde axon tracer techniques (Bentivoglio et al. 1980; Keizer et al. 1983; Katz et al. 1984; Katz and Iarovici 1990; Nance and Burns 1990), intracellular labeling procedures (Buhl and Lubke 1989), the recording of single cortical neuron responses in the macaque during the execution of complex manual tasks (Evarts 1966; Evarts et al. 1984; Mountcastle et al. 1975; Porter 1972), and using noninvasive techniques in human subjects to map those cortical areas that are activated when comparable manual tasks are executed (Deiber et al. 1991; Grafton et al. 1992; Raichle et al. 1994), the anatomical specification of corticospinal connections has developed quite rapidly in recent years.

The recognition that *several* somatotopically organized corticospinal neuron populations, each originating from a distinctive area of cortex and terminating on particular neuron populations in every segment of the spinal cord, constitute the direct corticospinal link in primates was an important step (Toyoshima and Sakai 1982; Murray and Coulter 1981; Tanji and Kurata 1983, 1985; Hutchins et al. 1988; Dum and Strick 1991; Nudo and Masterton 1990; Luppino et al. 1991; Galea and Darian-Smith 1994; Tanji 1994). Each of these widely distributed cortical neuron populations has its own distinctive connections and processes different sensorimotor and cognitive information. However, each such population has direct access to the spinal circuitry that mediates motoneuron activity and thus may modify the overall pattern of movement (Fetz et al. 1976; Asanuma et al. 1979; Shinoda et al. 1979; Kuypers 1981). These direct corticospinal projections constitute one series of parallel processing channels that must operate cooperatively. There is a second grouping of parallel corticospinal pathways (Kuypers 1981) which can, in principle, mediate the transfer of information from the cortex to the spinal cord, but their anatomy suggests that they function in a somewhat different mode to the direct pathways. Features common to each of these *indirect* corticospinal projections include the following: (a) they are multisynaptic, (b) the

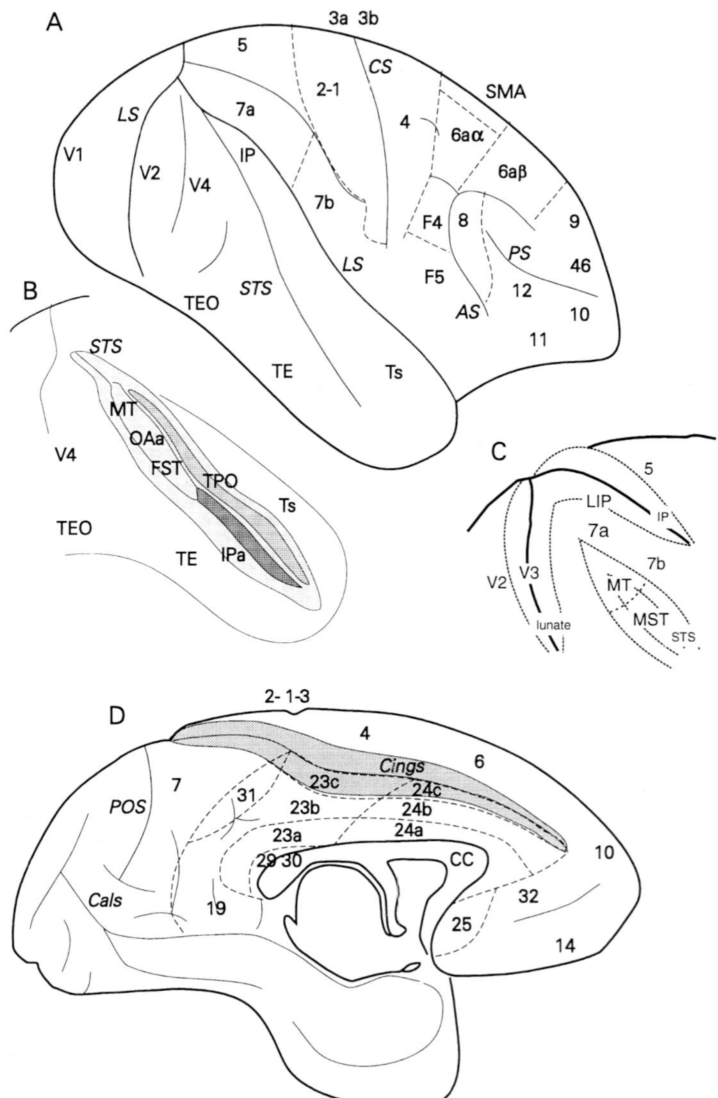

Fig. 2.1A–D. Topography of cytoarchitectonic areas of the cerebral cortex of the macaque monkey. This is a composite of several different maps, since no single complete map (e.g., Brodmann 1909; Vogt and Vogt 1919; von Bonin and Bailey 1947) fully corresponds to modern maps, which usually are concerned only with a selected region of cortex. All the indicated boundaries are approximate, as is so with any cytoarchitectonic map projected onto the intact, unfolded cerebral cortex. Each map shown is adapted from published maps; the following references are those actually used to prepare this figure. **A,B** Prefrontal cortex (Walker 1938). Frontoparietal cortex (Vogt and Vogt 1919; Matelli et al. 1986, 1989, 1991). Occipital and temporal cortex (Seltzer and Pandya 1989, 1994). **C** Visual cortex (Stein 1992). **D** Mesial cortex (Vogt et al. 1987a,b). *CS*, central sulcus; *IP*, intraparietal sulcus; *LS*, lunate sulcus; *AS*, arcuate sulcus; *PS*, principal sulcus; *LS*, lateral sulcus; *POS*, parieto-occipital sulcus; *Cals*, calcarine sulcus; *STS*, superior temporal sulcus; *SMA,* supplementary motor area; *Cings*, cingulate sulcus; *CC*, corpus callosum; *V1*, *V2*, *V3*, and *V4* are visual cortices. Remaining labels are of architectonic areas. In **B–D**, some sulci have been unfolded. *FST*, *TEO*, *TE*, *TPO*, *OAa*, divisions of the inferotemporal cortex; *MT*, division of the superior temporal cortex; *LIP*, lateral intraparietal cortex

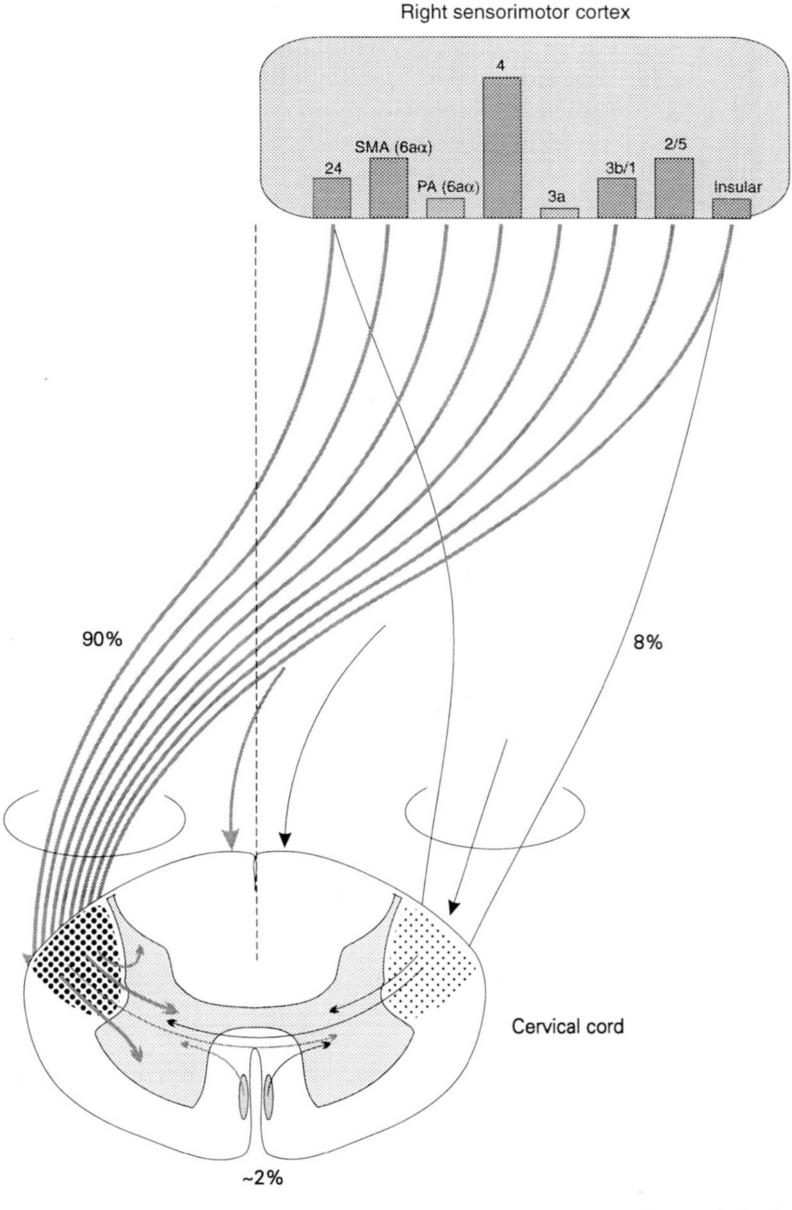

Fig. 2.2. Direct contralateral and ipsilateral corticospinal projections to the cervical spinal cord in the macaque. The multiple cortical origins from the frontal, parietal, medial, and insular cortex are represented in terms of their relative magnitudes. About 90% of this total corticospinal projection is contralateral and descends within the dorsolateral column, and 8% is ipsilateral within the same column. Approximately 2% of corticospinal fibers descend in the ventromedial column in the macaque (see text). Fibers in each of these descending pathways terminate mainly on the side of the cord in which they descend, but each also includes fibers which cross the midline to synapse in the intermediate zone. This implies that the cord below a hemisection is never fully isolated from direct corticospinal input. *SMA*, supplementary motor area. *PA*, poastarcuate cortex

bulbospinal component of each projection originates from a quite localized soma grouping in some part of the brain stem, and (c) each bulbospinal component has extensive cerebellar connections. The rubrospinal, tectospinal, vestibulospinal and several reticulospinal neuron populations operate as the spinal links in these indirect corticospinal projections.

We first briefly describe what is known of the structural organization of corticospinal neuronal connections in the macaque. The more immediate issues concerning *structural* organization that can be addressed by using these new methods include the following:

- The topography of all direct corticospinal neuron populations, including their cortical origins, trajectories and synaptic connections, and the spinal circuitry mediating the transfer of information from the cortex to motoneuron populations (Figs. 2.1, 2.2).
- The topography, circuitry, and synaptic organization of the neuronal populations that constitute the corticobulbospinal projections (see Fig. 2.12).
- The morphological characteristics, including the soma/dendrite organization of the constituent neurons at different levels within each corticospinal pathway (see Fig. 2.10).
- The cortical connections and inputs from subcortical neuron populations to each corticospinal neuron population (see Chaps. 3, 4).
- The postnatal maturational changes that occur in the different corticospinal neuron populations and their correlation with the acquisition of manual dexterity over this same period.
- The structural reorganization of the different corticospinal neuron populations following their transection and how these changes correlate with the rapid recovery of manual dexterity. In this review only passing reference is given to recent studies of the functional anatomy of corticospinal lesions.

As will be seen in the following pages, the systematic analysis of each of the above issues is quite incomplete to varying degrees. We address both new findings and some deficits in our current knowledge of most of these topics.

Although the model of the various corticospinal and corticobulbospinal pathways in the primate has changed substantially in the last 15 years, Kuypers' systematic review (1981) of the comparative anatomy of these "descending pathways" in mammals remains a landmark. Since his review, many details have been modified, often resulting from the application of tracing procedures that Kuypers himself developed, but the thoughtful generalizations that he incorporated into his synthesis have greatly influenced subsequent research.

Kuypers' emphasis was on the patterns of termination of the direct and indirect corticospinal projections within the spinal cord of different mammalian species. Two groups of bulbospinal (indirect corticospinal) projections were identified. In primates, the first group includes the vestibulospinal pathways and reticulospinal projections from the medial tegmental fields in medulla, pons, and midbrain. These projections are bilateral, located within the ventral and ventrolateral spinal fasciculi, and terminate mainly within the ventromedial part of the spinal intermediate zone at all segmental levels. The cortical inputs to these bulbospinal neuron populations originate mainly from localized cortical areas in the lateral part of areas 3a, 2 and the granular insular and retroinsular areas (for vestibulospinal projections, see Akbarian et al. 1992; Fredrickson and Rubin 1986; for reticulospinal projections, see Grantyn et al. 1993, Thielert and Thier 1993). The second group of indirect corticospinal projections include the corticorubrospinal (Holstege et al. 1988; Humphrey et al. 1984) and corticopontospinal (from the ventrolateral pontine tegmentum) projections (Keifer and Houk 1994). These projections are predominantly contralateral, descend in the spinal dorsolateral fasciculus, and terminate mainly but not solely in the lateral part of the spinal intermediate zone. The cortical input to the rubrospinal neuron population originates largely from precentral areas 4 and 6, and to a lesser extent from prefrontal area 8 and parietal area 5. As is illustrated later in this review, the direct corticospinal neuron populations originate from a much denser and more extensive area of cortex than the indirect projections, with corticospinal neuron populations in areas 4, 6aα (supplementary motor area, SMA; postarcuate cortex), the anterior cingulate cortex (area 24), areas 2 and 5, to a lesser extent areas 3a, 3b, and 1, and the insular cortex. The full topography of these various corticospinal and corticobulbar projections only became apparent with the use of the new retrograde and anterograde techniques during the 1980s and did not feature in Kuypers' synthesis. Rather, he emphasized the intrasegmental distributions of the spinal projections of the various direct and indirect corticospinal neuron populations and selected the corticomotoneuronal projections as having a special functional relevance in the execution of discriminating manual tasks. These issues are examined as they relate to the primate brain in the following sections. Recent views of the comparative anatomy of these pathways are summarized elsewhere by ten Donkelaar (1988).

In the above description, the bulbospinal projections are viewed as the spinal link of *indirect* corticospinal projections. Corticobulbar projections to the tectum, red nucleus, and the vestibular and brain stem reticular nuclei have in fact been identified, but it may well be simplistic to consider that these constitute a functional corticospinal complex. Few corticorubral axons, for example, synapse on rubrospinal neurons (see Fig. 2.12). Rather, most of these axons terminate on a neuron subpopulation in the red nucleus (Humphrey et al. 1984), which in turn projects to the inferior olive/cerebellar cortex/interpositus nucleus and finally gains indirect access to the rubrospinal neuron population. A corticorubrospinal projection of this complexity suggests that the cortical input to the red nucleus may have a quite limited access to the spinal cord and be greatly modified by its processing in the cerebellum. Similarly, the corticovestibulospinal projections are complex (Akbarian et al. 1992, 1994) and may have a mainly visuomotor role.

2.2 New Methods

The axon tracer procedures which have been developed in the last 15 years (retrograde tracers; Bentivoglio et al. 1980; Keizer et al. 1983; Katz et al. 1984; Katz and Iarovici 1990; Darian-Smith et al. 1990a; Galea and Darian-Smith 1994; Nance and Burns 1990; Chang 1993; Rajakumar et al. 1993) provide a more inclusive and more quantitative representation of these corticospinal projections than was possible in the earlier investigations. Tracer procedures in which the

Table 2.1. Neuron labeling procedures

Tracer	Processing and display	Transport time (corticospinal)	Comment
Retrograde tracers			
Fast blue (FB) 2% aqueous	Fluorescent filter-λ, 360 nm	10–14 days	Necrotic zone in injection zone necessary for labeling cytoplasm labeled blue
Diamidino yellow (DY) 2% aqueous	Fluorescent filter-λ, 360 nm	10–14 days	Necrotic zone in injection zone necessary for labeling nucleus labeled greenish yellow
Rhodamine latex microspheres (RLM) in suspension as supplied	Fluorescent filter-λ, 550 nm	10–14 days	Reddish gold beads in cytoplasm; in situ for several months; robust enough to withstand some histochemical processing
Green latex microspheres (GLM) in suspension as supplied	Fluorescent filter-λ, 450 nm	10–14 days	Fluorescein green beads in cytoplasm
Wheat-germ agglutinin (WGA-HRP) 2% aqueous		2–5 days	Better labeling obtained with minimum fixation; also anterograde, but boutons and fibers difficult to differentiate
Anterograde tracers			
WGA-HRP (see above)			
Biotinylated dextran amine (BDA; MW, 10000); 10%–15% aqueous	Immunohistochemistry; light microscopy: black labeling	30 days for corticocervical transport	Labeling of fibers and boutons
Lucifer yellow dextran (LYD, MW, 10000); 10%–15% aqueous	Immunohistochemistry; light microscopy: brown labeling	30 days for corticocervical transport	Terminal labeling is good, BDA is better
Tetramethyl-rhodamine dextran (TMR; MW, 10000); 10% aqueous	Fluorescence microscopy; filter-λ, 550 nm	30 days for corticocervical transport	Good labeling of fibers and terminals, but boutons difficult to distinguish
Fluorescein dextran (FD; MW, 3000); 10% aqueous	Fluorescence microscopy; filter-λ, 450 nm	30 days for corticocervical transport	Both anterograde and retrograde labeling

MW, molecular weight.

label is transported either anterogradely or retrogradely from the zone of uptake have been used in several laboratories to examine these corticospinal projections in primate species (Biber et al. 1978; Coulter and Jones 1977; Murray and Coulter 1981; Toyoshima and Sakai 1981; Ralston and Ralston 1985; Martino and Strick

1987; Hutchins et al. 1988; Keizer and Kuypers 1989; Nudo and Masterton 1990; Dum and Strick 1991; Bortoff and Strick 1993; Galea and Darian-Smith 1994, 1995). The labels used in our laboratory, with brief comments on their advantages and limitations, are listed in Table 2.1.

Retrograde labeling is the current technical choice for examining the widespread neuron soma distributions of corticospinal neuron populations. Following the injection of a retrogradely transported label (e.g., horseradish peroxidase, HRP, or the fluorescent tracer fast blue, FB; see Table 2.1) into a particular spinal cord segment, these corticospinal somas can be visualized (Murray and Coulter 1981; Toyoshima and Sakai 1982; Hutchins et al. 1988) and their regional densities mapped and correlated with the landmarks and cytoarchitecture of the cerebral cortex (Dum and Strick 1989, 1991, 1992; Galea and Darian-Smith 1994, 1995). A library of distinctive retrograde labels that can be used simultaneously in one monkey has been developed in recent years; this not only reduces the number of animals used in a particular study, but also greatly increases the precision with which the spatial distributions of different labeled neuron populations can be compared (Bentivoglio and Molinari 1984; Darian-Smith et al. 1990a). The technical limitations of retrograde labeling lie in the specification of the zone of uptake in the spinal cord of the retrogradely transported label by axons and their terminals (Payne and Peace 1989; Condé 1987; Darian-Smith et al. 1990a; Galea and Darian-Smith 1994, 1995). Most modern retrogradely transported labels are taken up by both axons and their terminals. These limitations can be resolved to a great extent, as will be demonstrated in this chapter, so that recent studies of the corticospinal soma distributions in the macaque have visualized an unforeseen regional complexity.

Anterograde labeling has been extensively used to visualize the axonal projections and terminations in the spinal cord of corticospinal neurons. The cortical injection of a tracer, such as wheat-germ agglutinin (WGA-HRP) (Cheema et al. 1984; Ralston and Ralston 1985) or tritiated amino acids (Coulter and Jones 1977), labels many but not all of the terminal projections in the spinal cord of those cortiocospinal tract fiber cells whose somas are located within the localized injection site. Until 2–3 years ago, either HRP or autoradiography was the only practical anterograde label for these studies, which meant that the total return from each monkey used was only one or sometimes two projections from selected cortical areas. This limitation has to some extent been overcome by the recent introduction of biotinylated dextran amine (BDA; molecular weight, 10000), lucifer yellow dextran (LYD), tetramethyl-rhodamine dextran (TMR), and fluorescein dextran (FD), since these can be used as separate injections in one monkey, along with retrograde fluorescent tracers. However, only a few studies of the corticospinal projections have yet exploited this advance. While this new anterograde labeling is useful for defining corticospinal terminal branchings and endings, it is a rather clumsy procedure for analyzing the widely distributed *origins* of the corticospinal neuron populations, because of (a) the many labeling injections still needed for any extensive study and (b) the difficulty in assessing and comparing the densities of the projection labeled by any two injections of an anterograde label (e.g., see Bortoff and Strick 1993). Combining the new anterograde and retrograde labeling of each neuron population is really needed to obtain the best resolution of its projections (see Chap. 3).

The *intracellular injection of a label* into individual corticospinal neurons is yet another way of visualizing their morphology. This procedure is especially valuable for examining the morphology of the dendritic tree, dendritic spine distributions, the origin of the axon from the soma and its intracortical collateral branching, and, still largely for the future, the distribution of synapses of known origin on the cell's soma and dendritic branches. Early attempts with intracellular labeling of corticospinal neurons in the macaque (Lawrence et al. 1985; Ghosh and Porter 1988a,b) were in vivo, with corticospinal cells being identified by their antidromic discharge following downstream electrical stimulation. Such heroic experiments yielded only a few successfully injected neurons, which necessarily restricted the questions that could be asked concerning the morphology of individual corticospinal neurons in the macaque. More recently, the use of fixed cortical slices for injecting prelabeled corticospinal neurons has transformed this approach (for methods, see Buhl and Lubke 1989; Freire 1990; Larkman 1991a–c). Table 2.1 gives brief technical details of the intracellular labels that we have used so far, and Figs. 2.10 and 3.11 illustrate some of the results.

Three other technical advances that have shaped much of the recent cortical research in the primate are the following:

1. The coincident analysis of the responses of single neuron in a particular area of cortex and the performance of a carefully defined manual task by a macaque (Evarts 1964, 1966; Evarts et al. 1984; Mountcastle 1975; Mountcastle et al. 1975; Porter 1972; Porter and Lewis 1975; Hyvärinen 1982; Georgopoulos 1986, 1991; Darian-Smith et al. 1984, 1985)
2. The use of noninvasive procedures for measuring regional neuronal activity in the human cerebral cortex, again during the execution of particular sensorimotor manual tasks (PET and functional MRI; Zeki et al. 1991; Corbetta et al. 1993; Seitz et al. 1993; Posner 1993; Burton et al. 1993; Raichle et al. 1994)
3. The noninvasive use of electromagnetic stimulation to excite, either directly or indirectly, particular corticospinal neuron populations in the human or monkey brain (Brouwer and Ashby 1992; Baker et al. 1994; Flament et al. 1992a,b; Vanderlinden and Bruggeman 1993)

2.2.1 Mapping the Distributions of Labeled Corticospinal Neuron Somas

The widespread use of multiple fluorescent labels in one experimental monkey, combined with a computer-linked digitizing system for mapping the locations of retrogradely labeled fluorescent somas in serial histological sections (e.g. Galea

Fig. 2.3a–n. Series of 14 coronal maps of the macaque cortex (selected from 42) of the four distributions of corticospinal neuron somas labeled with the retrogradely transported fluorescent dyes fast blue (*FB*), diamidino yellow (*DY*), rhodamine latex microspheres (*RLM*), and green latex microspheres (*GLM*). Each dye has been injected into a particular site in the cervical spinal cord (C8), as shown in Fig. 2.5. The location of each coronal map is shown in the *inset* of the forebrain. *Ci.S*, cingulate sulcus; *PS*, principal sulcus; *SAS*, superior arcuate sulcus; *IAS*, inferior arcuate sulcus; *AS*, arcuate sulcus; *LS*, lateral sulcus; *CS*, central sulcus; *IPS*, intraparietal sulcus ▶

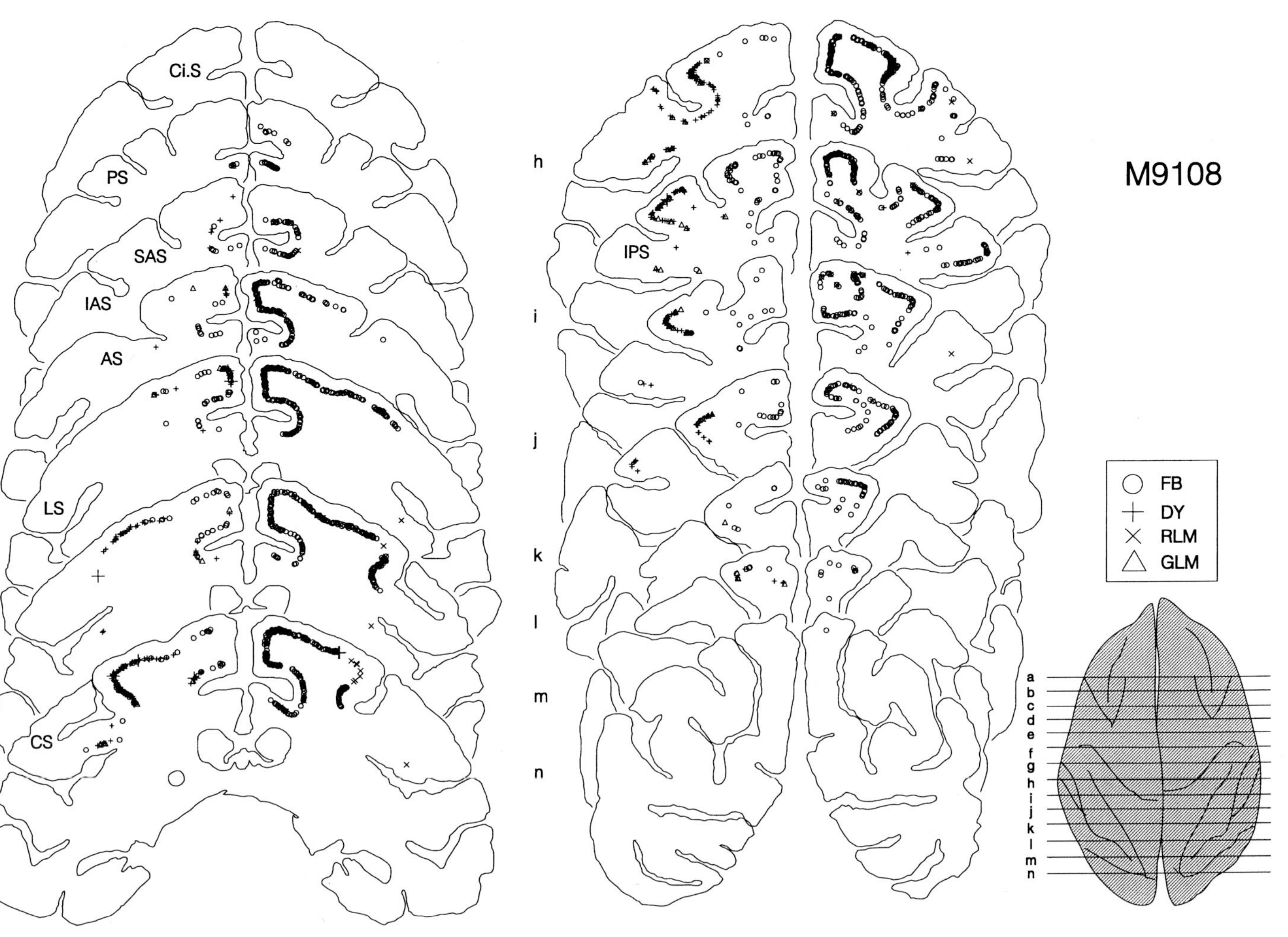
M9108
FB
DY
RLM
GLM
a
b
c
d
e
f
g
h
i
j
k
l
m
n
Ci.S
PS
SAS
IAS
AS
LS
CS
IPS

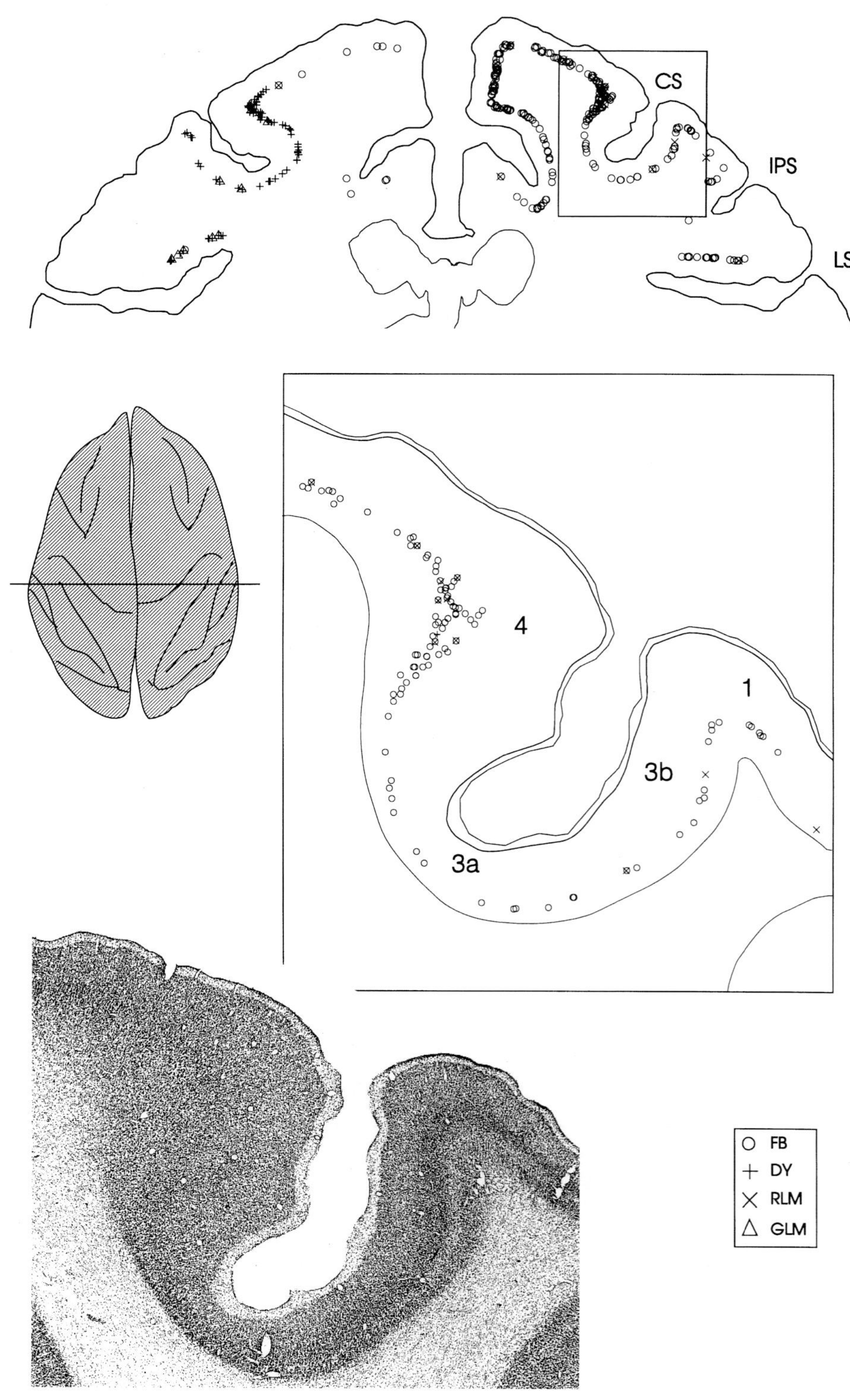
CS
IPS
LS
4
1
3b
3a
FB
DY
RLM
GLM

and Darian-Smith 1994), has highlighted the need for reliable maps which illustrate these soma distributions at the necessary resolution. Since we use these maps extensively to illustrate this review, it may be useful to know how they are constructed and what their uses and limitations are. Figure 2.3 illustrates some of a series of 49 coronal maps of the labeled corticospinal soma distributions of neurons following the injection of four fluorescent dyes (FB; diamidino yellow, DY; rhodamine latex microspheres, RLM; and green latex microspheres, GLM) into different sites in the cervical spinal cord (see legend). These maps were prepared using 50-μm sections selected from a series of serial sections of the hemispheres. Figure 2.4 is an enlarged view of section *h* in Fig. 2.3, illustrating the soma distribution around the central sulcus and its relations to the cortical cytoarchitecture. All the labeled somas lie within lamina V, regardless of the cortical area (areas 4, 3a, 3b, and 1). The spatial resolution of any maps of this soma distribution depends on the measurements made within each coronal map, and with three-dimensional maps the interval between successive maps becomes important. The scale of the map is selected to provide the level of detail required for the planned analysis.

A series of coronal maps, as in Fig. 2.3, describes the three-dimensional distribution of each population of labeled corticospinal somas. However, it is quite difficult to appreciate the spatial complexity of this distribution and compare it quantitatively with maps of neurons labeled with other dyes without further graphical analysis. Advantage can be taken of the fact that the corticospinal soma distribution within lamina V of the cerebral cortex constitutes a monolayer. This layer can be unfolded and then projected onto a planar surface, and the soma distribution plotted as a contour map (as in Fig. 2.5) or in three dimensions (Fig. 2.6). The problems associated with the representation of a curved, folded surface in a geometrically simple form are well known (Robinson et al. 1984; Monmonnier 1991). Thus no flat projection of a folded, curved surface can accurately represent all its features: distances, areas, shapes, angles, and directions will not all match precisely in the planar and original three-dimensional surface maps. The familiar Mercator map of the global Earth illustrates this. The best map to use is that which least distorts those features which are immediately relevant. In fact, in assessing the distributions of labeled neurons in the highly convoluted cortex of the macaque, planar maps are best read in conjunction with the original coronal maps.

The practical details of preparing these contour maps are described elsewhere (Darian-Smith et al. 1993; Galea and Darian-Smith 1994, 1995). For the soma distributions considered in this review, coronal maps prepared at 1- or 2-mm intervals were quite adequate to illustrate the features discussed: between 900 and 2000 soma density estimates were used in constructing each map (e.g. in

◀ **Fig. 2.4.** Detail of map *h* from Fig. 2.3, with rostrocaudal location indicated in the *inset* of the forebrain, showing soma distributions of corticospinal neurons labeled with fast blue (*FB*) and rhodamine latex microspheres (*RLM*). Data from the same macaque is also shown in Fig. 2.5. Labeled somas are all within cortical lamina V. Marked difference in projection density in cortical area 4 (motor) and 3a (somatosensory). Histological section is stained with cresyl violet. *CS*, central sulcus; *IPS*, intraparietal sulcus; *LS*, lateral sulcus; *DY*, diamidino yellow; *GLM*, green latex microspheres

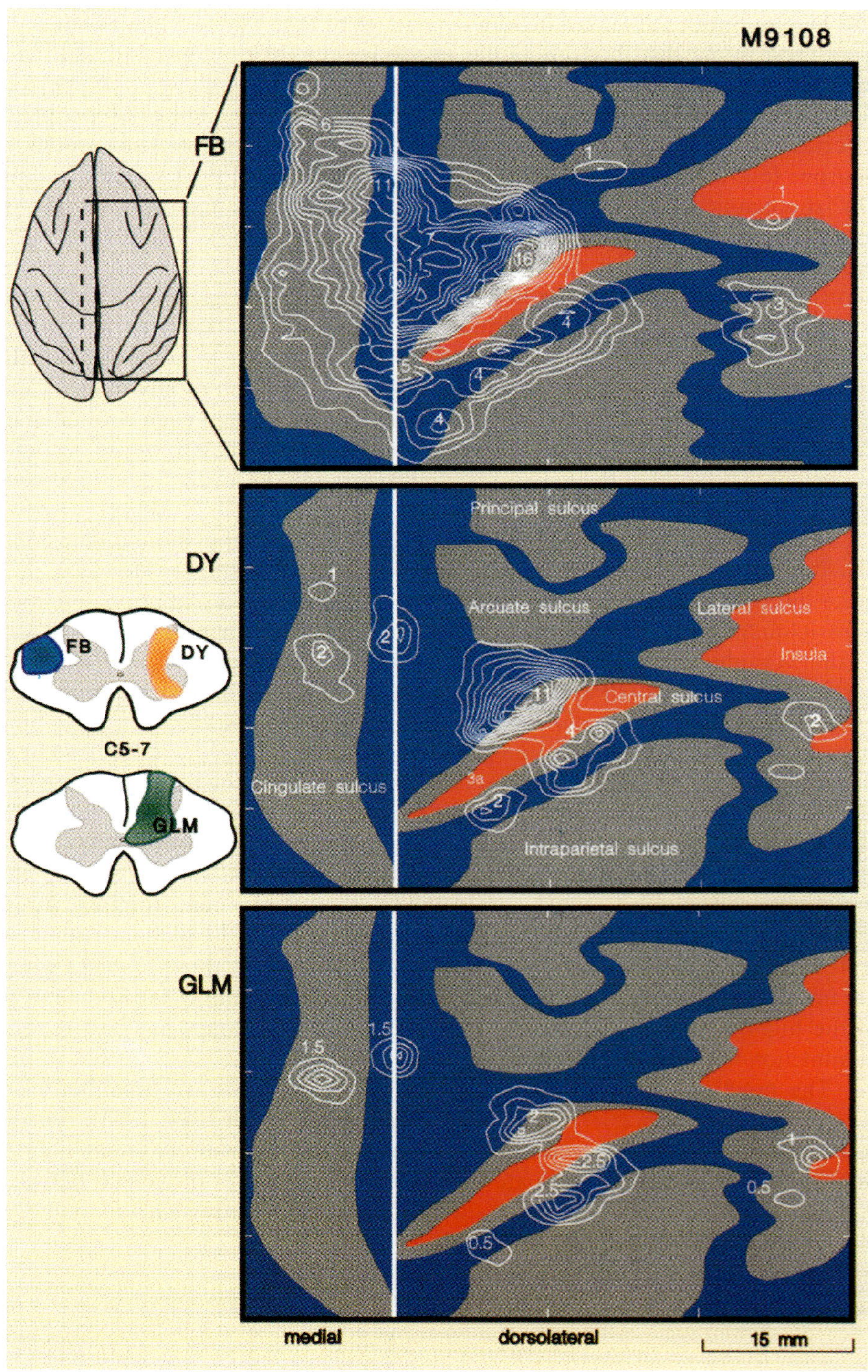

M9108
FB
DY
GLM
FB
DY
C5-7
GLM
Principal sulcus
Arcuate sulcus
Lateral sulcus
Insula
Central sulcus
Cingulate sulcus
3a
Intraparietal sulcus
medial
dorsolateral
15 mm

Figs. 2.7, 2.8). When greater resolution was needed (e.g. in locating corticospinal somas in lamina V or where precipitate changes in soma density occurred), the original coronal maps were sometimes supplemented with additional coronal maps contructed at intervals of 500 µm or less; these were examined individually and also incorporated into the data base on which the contour maps were based.

The same data base used to construct each contour map also specifies the total areal extent of the soma distribution and the total number of somas counted, which with appropriate sampling procedures provides an estimate of the total number of corticospinal neurons labeled by the dye injection. These parameters have been useful in analyzing the postnatal maturation of corticospinal neuron populations in the macaque (Galea and Darian-Smith 1995) and are considered later in this chapter.

2.3 Direct Corticospinal Projections

2.3.1 Origins and Projections

Early in this century, Sherrington (1947) and Leyton and Sherrington (1917) concluded that the somatotopically organized corticospinal projection in the ape and macaque originates from the primary precentral motor cortex (Brodmann's area 4; Fig. 2.1), descends through the internal capsule and brain stem, largely decussates in the medulla to form the pyramid, and descends in the spinal cord to terminate in particular segments. The executive commands that determine voluntary action were thought to be transmitted by this pathway and to regulate movement patterns by modulating activity in local spinal reflex pathways. Subsequent to Sherrington's analysis, it was gradually appreciated that direct corticospinal projections arise not only from the electrically excitable precentral cortex, but also from the somatosensory cortex (Woolsey et al. 1952; Russell and DeMeyer 1961), somatosensory area II (Toyoshima and Sakai 1981), the posterior parietal cortex (Coulter and Jones 1977), the SMA (Penfield and Boldrey 1937; Penfield and Welch 1951; Murray and Coulter 1981; Toyoshima and Sakai 1981; MacPherson et al. 1982; Tanji 1994), and, as shown recently, also from the anterior cingulate cortex (Murray and Coulter 1981; Hutchins et al. 1988; Shima et al. 1991; Dum and Strick 1991; Luppino et al. 1991; Matsuzaka et al. 1992; Galea and Darian-Smith 1994, 1995).

Figure 2.5 illustrates the distributions of corticospinal neuron somas retrogradely labeled following the injection of FB, DY, and GLM into different

◀ **Fig. 2.5.** Contour maps of distributions of somas of corticospinal neurons labeled with three retrogradely transported dyes (fast blue, *FB*; diamidino yellow, *DY*; green latex microspheres, *GLM*), injected into the different sites in spinal C5–7 shown in the *inset* on the *left* in the macaque. Each distribution has been plotted as a planar projection onto the unfolded lamina V of the sensorimotor cortex *contralateral* to the dye injection site. *Inset cortex* illustrates the region that has been unfolded and flattened. See text for details. *Vertical white line* is the sagittal midline, with the mesial cortex unfolded to the left, and the dorsolateral cortex to the right. Sulci are *gray*, the gyral surface is *blue*, and area 3a is *red*

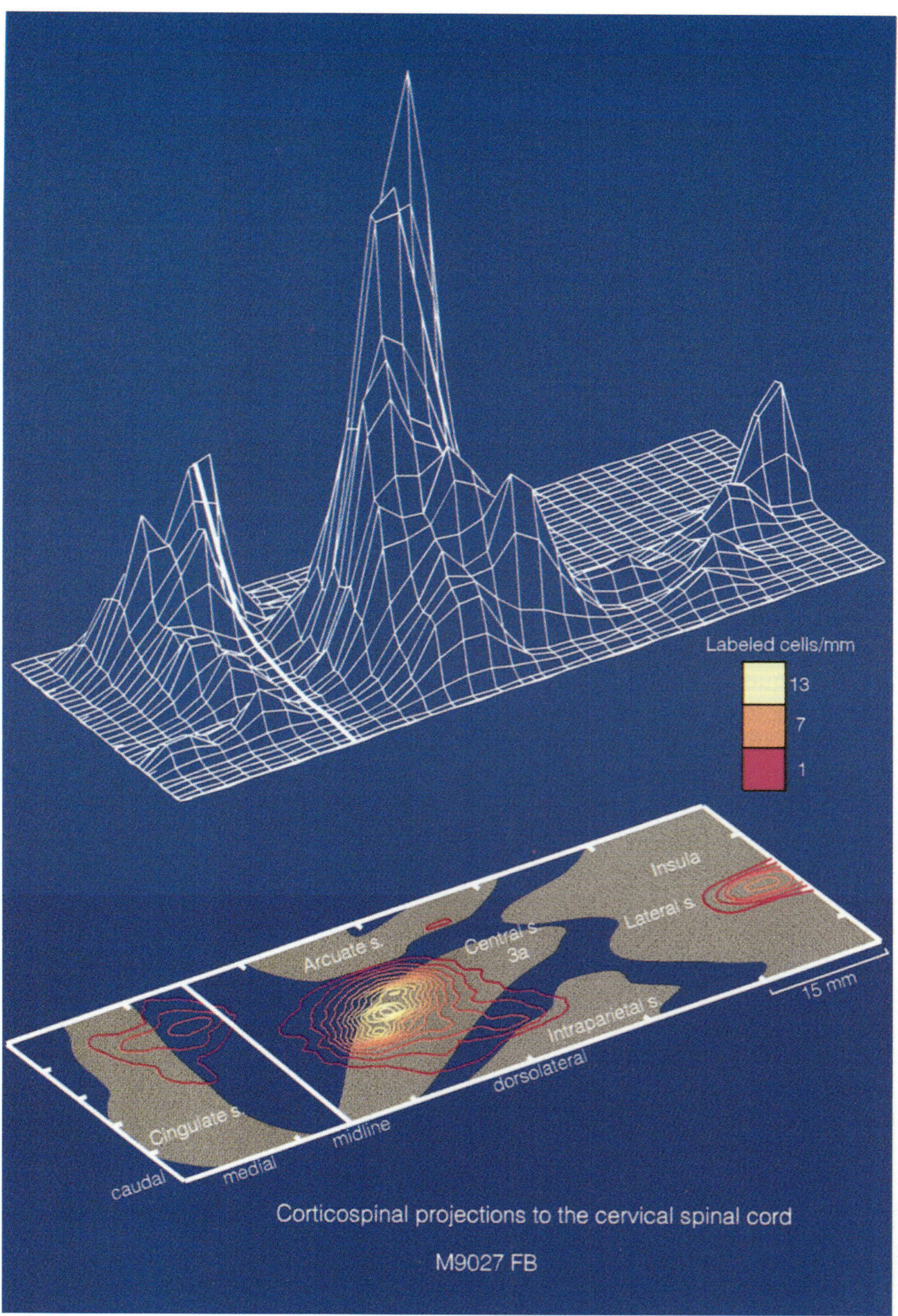

Fig. 2.6. Three-dimensional (*top*) and two-dimensional (*bottom*) contour display of the distribution of somas of corticospinal neurons labeled following the injection of retrogradely transported fluorescent label (fast blue, FB) in spinal segment C5 of a mature macaque. The uptake zone of the label was in the intermediate zone and part of the dorsal and ventral horns, but not in the white matter. Each density map is plotted onto the unfolded lamina V of the cortex, with the mesial cortex to the left of the midline, and the dorsolateral cortex to the right. Unfolded sulci are shaded *grey*. Multiple, separate corticospinal projections originate from the motor area 4 (hand/forearm representation), anterior cingulate area, supplementary motor area (SMA), area 2/5, and the insular cortex

regions in the spinal segments C5–7 in the macaque. The corticospinal soma distribution map at the top of Fig. 2.5 is the most complex of the sequence shown and illustrates the cortical origins of all but a few percent of corticospinal neurons whose axons traverse or terminate in the contralateral cervical spinal cord. The retrograde tracer used (FB; see Table 2.1) was injected into C5 and had spread throughout the dorsolateral column, within which most corticospinal fibers are located in the macaque (Leyton and Sherrington 1917; Kuypers 1981; Ralston and Ralston 1985; Bortoff and Strick 1993; Galea and Darian-Smith 1994, 1995). FB and most other commonly used retrogradely transported fluorescent tracers (DY, RLM, and GLM) are taken up at the injection site by both the axon terminals and the axon shafts (Darian-Smith et al. 1990a; Galea and Darian-Smith 1994). This accounts for the extensive distributions of labeled somas of the different corticospinal neuron populations. Contrasting with this extensive distribution of FB-labeled cells were the localized distributions of soma populations labeled with DY and GLM (Fig. 2.5). The "uptake zones" were also in C5–7, but were restricted to the gray matter with DY and to the dorsal column, dorsal horn, and intermediate zone with GLM. Since there are no corticospinal fibers in the macaque's dorsal column (Kuypers 1981; Galea and Darian-Smith 1994) and only the terminals of corticospinal axons in the spinal gray matter, the distributions of labeled somas in the middle and lower contour maps of Fig. 2.5 were solely of corticospinal neurons projecting to the contralateral side of cervical segments C5–7. Even so, these more restricted projections to only one or two cervical spinal segments originate from several discrete areas, with peaks in motor area 4, the SMA, and the anterior cingulate, postcentral, and peri-insular cortex. Figure 2.6 illustrates in three dimensions these separate corticospinal populations projecting to the gray matter of C5 in another monkey, but here the local "densities" of the projections are plotted on the z-axis and highlight the discreteness of several of the projections from primary motor cortex (area 4), the SMA, the anterior cingulate cortex, parietal areas 2 and 5, and the insular cortex, including somatosensory area SII. These three-dimensional maps also illustrate the sparse corticospinal projections from somatosensory areas 3a and 3b.

The contour maps in Fig. 2.7a illustrate the contralateral origins of corticospinal fibers terminating in spinal segments C5, T2, L2–3. Those in Fig. 2.7b show contralateral origins of fibers traversing the spinal cord at levels C5–6, T12, and S3. These maps illustrate the somatotopy of the precentral primary motor cortex, which Ferrier (1876) plotted out 120 years ago using localized electrical stimulation of this cortex. However, this retrograde labeling of neurons also demonstrates several other corticospinal projections to the full rostrocaudal extent of the spinal cord which originate from the anterior cingulate cortex, the SMA, parietal areas 2 and 5, and the insular cortex, respectively. The spatial resolution achieved with this sequence of maps in different monkeys was insufficient to demonstrate a somatotopic pattern among the somas of corticospinal neurons projecting to the different spinal segments. Other studies support the view that the SMA is somatotopically organized (Tanji 1994), but the evidence for a somatotopic organization of the cingulate corticospinal projections is still equivocal (Luppino et al. 1991, 1994).

In Fig. 2.8 the cortical origins of the contralateral and *ipsilateral* corticospinal projections to the macaque's spinal cord (C2 and more caudally) are compared.

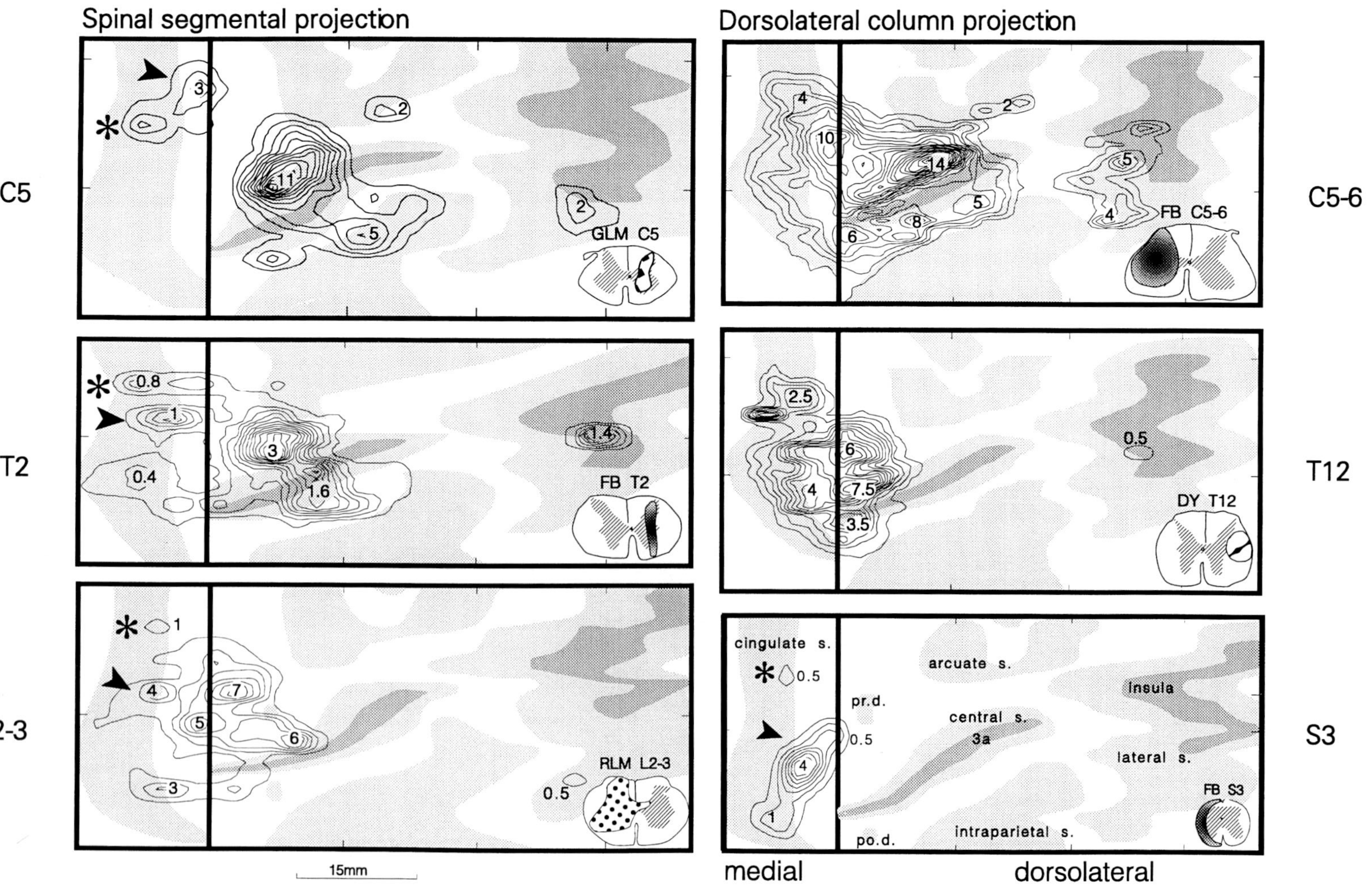

Spinal segmental projection
Dorsolateral column projection
C5
T2
L2-3
C5-6
T12
S3
GLM C5
FB T2
RLM L2-3
FB C5-6
DY T12
FB S3
15mm
cingulate s.
pr.d.
po.d.
arcuate s.
central s.
3a
intraparietal s.
insula
lateral s.
medial
dorsolateral

Although the ipsilateral projection, with axons located in the spinal dorsolateral column, constitutes only approximately 8% of the total corticospinal projection in the mature macaque (Toyoshima and Sakai 1981; Galea and Darian-Smith 1994, 1995), discrete projections matching with the contralateral projections from the different frontal, parietal, and insular areas were observed. Of course, cataloguing corticospinal fibers as contralateral or ipsilateral on the basis of retrograde labeling simply means that the fibers in the zone of uptake of the retrograde label originate from the cortex contralateral or ipsilateral to the injection in the cervical spinal cord. Determining whether these axons or their terminals cross the midline at some point rostral or caudal to the labeling site requires additional visualization, best done with anterogradely transported labels. We have recently shown (M.P. Galea and I. Darian-Smith, unpublished) with anterograde labeling that terminal branches of some of the so-called "ipsilateral" fibers in the dorsolateral column in the macaque's cervical cord cross in the isthmus below the spinal canal and project to the contralateral intermediate zone (see Figs. 2.2 and 2.11 for details). This crossover of corticospinal fibers has proven to be significant in the recovery of manual dexterity following focal spinal lesions (Ralston and Ralston 1985; Galea and Darian-Smith 1995). Anterograde labeling of corticospinal neurons originating from area 4, the SMA, and parietal area 5 also visualizes a small projection of axons descending in the ipsilateral dorsolateral column and terminating in the ipsilateral intermediate zone and the ventral horn in the cervical spinal cord (i.e., "true" ipsilateral corticospinal neurons). Wassermann et al. (1994) have used transcranial electromagnetic stimulation to demonstrate a similar ipsilateral corticospinal projection in normal human subjects. Whether the ipsilateral corticospinal projections that can be identified in human subjects following an intracapsular stroke (Bastings et al. 1995) and whether those in the rare familial condition of mirror movements of the limbs (Cincotta et al. 1994) are the result of the disorder or simply match with the ipsilateral fibers found in the normal subject remains to be demonstrated.

The fractions of the different corticospinal projections to the contralateral spinal cord are listed in Table 2.2 (See also Fig. 2.9) and are similar to other recent descriptions (Toyoshima and Sakai 1981; Sessle and Wiesendanger 1982; Nudo and Masterton 1990). A sharp falloff in the corticospinal projection from areas 3a and 3b/1 of the postcentral cortex, with a denser projection from parietal areas 2 and 5, is also seen. Note that the above description relates to corticospinal fibers traversing the spinal dorsolateral column in the macaque. Fibers in the ventromedial column probably account for only approximately 1% of the total corticospinal population in these monkeys. Only recently have we identified to

◀ **Fig. 2.7.** Contour maps of soma distributions of corticospinal neurons projecting to spinal segments C5, T2, and L2–3, respectively (*left*). Projections mapped are to the contralateral spinal cord and illustrate the somatotopy of these projections and their multiple cortical origins. *Asterisks* and *arrowheads* indicate anterior cingulate and supplementary motor area (SMA) projections. The three contour maps on the *right* illustrate the cortical origins of *all* corticospinal axons in the dorsolateral fasciculus at C5–6, T12, and S3, respectively. *GLM*, green latex microspheres; *FB*, fast blue; *RLM*, rhodamine latex microspheres; *DY*, diamidino yellow; *s.*, sulcus; *pr.d.*, precentral dimple; *po.d.*, postcentral dimple

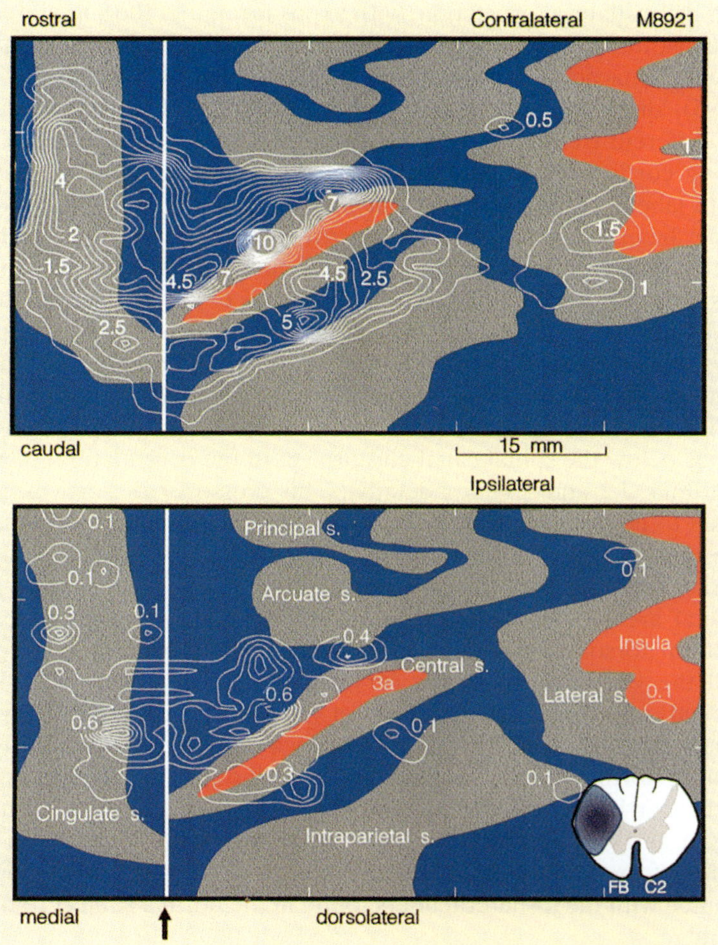

Fig. 2.8. Contour maps of the soma distributions of corticospinal neurons projecting to the contralateral and ipsilateral cervical spinal cord (C2) in the macaque. Map of left hemisphere is plotted as a mirror image to simplify the comparison of the maps. The retrograde label was fast blue (FB), injected at the site shown in the *inset*. The isodensity lines are labeled, the *numbers* referring to the peak densities in each area. Cortical landmarks are coded as in Fig. 2.6. Although the estimated total ipsilateral projection was numerically only approximately 10% of the total projection, the relative projections from the medial, frontal, parietal, and insular cortices were similar. *s.*, sulcus

our satisfaction these ventromedial fibers in the cervical spinal cord of *M. fascicularis*, using the anterograde tracers LYD and BDA.

No estimates comparable to those done in the macaque of the labeled fractions of the corticospinal neuron populations originating from the different cortical fields and descending in the dorsolateral and ventromedial columns have been reported in the apes. However, degeneration of corticospinal fibers resulting from focal injury or motoneuron disease in humans (Nathan et al. 1990) or induced experimentally in chimpanzees and gorillas (Leyton and Sherrington 1917; Fulton 1949; Kuypers 1981) unequivocally labels the ventromedial column,

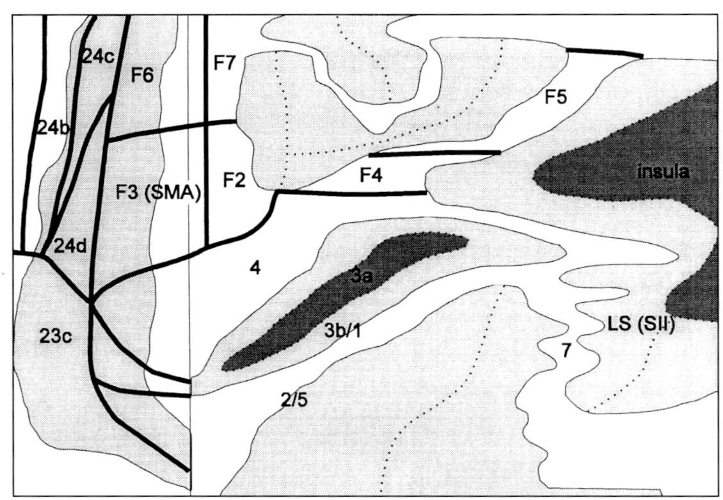

Fig. 2.9. Unfolded sensorimotor cerebral cortex of macaque monkey showing different areas in which somas of labeled corticospinal neurons were mapped (Galea and Darian-Smith 1994). *Light grey areas*, unfolded sulci: cingulate sulcus on the *left*; intraparietal sulcus *lower center*; lateral sulcus (*LS*) on the *right*; arcuate and principal sulci at the *top*; central sulcus includes 3a (*dark*) and parts of areas 3b/1 and 4. The abbreviations used are based on a combination of the classifications by Vogt and Vogt (1919), Matelli et al. (1991), and Galea and Darian-Smith (1994). This map illustrates the extent of each area in which labeled somas were located when constructing the data listed in Table 2.2, in which data from five "mature" and five newborn macaques were used. The mean and standard error of mean are given in Table 2.2

Table 2.2. Regional distributions of corticospinal somas in mature and newborn macaques (see also Fig. 2.9)

Cortical area	Mature macaque	Newborn macaque
24b ..	–	4.6 ± 1.4
24c ...	6.1 ± 1.2	3.9 ± 0.9
24d ..	–	3.6 ± 0.8
23	4 ± 1.2	3.6 ± 0.8
F6 (6aα)	1.3 ± 0.8	2.7 ± 1.4
F7	–	0.5 ± 0.3
F3 (SMA)	15.3 ± 2.7	10.6 ± 2.2
F2 (6aα)	5.8 ± 2.4	5.3 ± 0.9
F4	2.6 ± 2	2.9 ± 1.6
F5	0.3 ± 0.2	2.6 ± 1.1
4	35.4 ± 2.4	27.5 ± 4.6
3a	2.2 ± 0.9	1.4 ± 0.8
3b/l ..	9 ± 3.2	7.4 ± 2.1
2/5 ...	12.9 ± 1.4	10.4 ± 1.9
7	2.2 ± 0.9	5.2 ± 1.1
SII ...	2.3 ± 0.3	2.9 ± 0.6
Insular and peri-insular	1.1 ± 0.6	3.4 ± 1.9
Prefrontal	–	0.7 ± 0.5
Orbitofronal and opercular	–	1.2 ± 1.4

SMA, supplementary motor area.

implying a more substantial projection (perhaps 10% of the whole pyramidal projection; Fulton and Sheehan 1935) than that in the macaque. It has also been argued from clinical observations (Nathan et al. 1990) that substantial differences may occur in the proportionate distributions of the different corticospinal fiber populations in individual human brains.

2.3.2 Multiple Corticospinal Neuron Populations in the Macaque

The macaque's corticospinal neuron populations and their projections have now been experimentally examined periodically but intensively for about 100 years, and yet their functional anatomy continues to surprise us. For example, it was not until 1988 that the substantial projection from the anterior cingulate cortex, constituting about 6% of the whole corticospinal neuron population, although suggested by earlier studies (e.g., Murray and Coulter 1981), was more fully identified by Hutchins et al. and others (Shima et al. 1991; Mushiake et al. 1991; Matsuzaka et al. 1992). Axon tracer procedures with better resolution have in part accounted for this successive identification of new corticospinal neuron populations, but new methods developed for carefully matching and comparing the distributions of corticospinal neurons in the curved and folded sensorimotor cortex have also helped these studies. A review of the various maps of the distributions of the somas of corticospinal neurons in Figs. 2.5–2.8, 2.15, and 2.17 recalls the observed variations in these soma distributions, due in part to the labeling procedures, but also probably reflecting differences in individual monkeys. Many such maps must be analyzed to identify their invariant elements. The following paragraphs summarize the distinctive corticospinal neuron populations that have so far been reliably identified in the macaque and can be identified in the relevant figures. Each corticospinal neuron population has unique thalamic and cortical connections, which are considered in Chaps. 3 and 4, respectively.

2.3.2.1 Projections from the Frontal Cortex (Areas 4 and 6aα, Supplementary Motor Area, and Postarcuate Cortex)

The three separate somatotopically organized corticospinal neuron populations that can be identified in the precentral agranular areas 4 and 6aα are as follows:

1. The largest single population filling most of area 4, except in the lateral region of orofacial representation, with a sharp reduction in this projection caudally near the border between areas 4 and 3a, and with a more gradual fall off rostrally in area 6aα
2. The SMA in the medial part of area 6aα
3. A small neuron cluster in the lateral lip of the arcuate spur and the lateral arcuate sulcus

The population in area 4 in the macaque is somatopically organized. SMA also has a somatotopic motor representation, although this was not established without some confusion (Tanji 1994; Mitz and Wise 1987); the face, forelimb, and hindlimb motor representations are rostrocaudally arranged, as are the corticospinal projections to the cervical and lumbar spinal cord (Burman 1992). The postarcuate corticospinal neuron population projects mainly to the cervical spinal cord; its identity as a distinctive population is apparent from its spatial separation from other corticospinal neuron populations (Galea and Darian-Smith 1994).

2.3.2.2 Projections from the Cingulate Cortex (Areas 24 and 23)

The recent identification of corticospinal projections from the cingulate cortex (Murray and Coulter 1981; Hutchins et al. 1988; Dum and Strick 1991a,b; Shima et al. 1991; Matsuzaka et al. 1992; Galea and Darian-Smith 1994, 1995) has raised the question of whether separate populations, each with input from the limbic system, are located anteriorly within area 24 and more caudally in area 23. Both tracer studies and electrophysiological investigations, using microstimulation and the recording of single neuron responses (Shima et al. 1991; Luppino et al. 1991), imply that there are two such spatially separate populations, but their separate somatotopy and physiological differentiation have still to be resolved.

2.3.2.3 Projections from the Parietal and Insular Cortex

The sharp transition in the cytoarchitecture, neuron response characteristics, and behavioral deficits resulting from focal lesions at the boundary between cortical areas 4 (motor) and 3a (somatosensory) is mirrored in their corticospinal projections. As seen in the contour maps in Figs. 2.5–2.8, 2.15, and 2.17 and in Table 2.2, the densest and most extensive corticospinal projection in the macaque originates from primary motor area 4, but the density of this projection drops sharply at its caudal boundary with area 3a (Sessle and Wiesendanger 1982; Nudo and Masterton 1990; Galea and Darian-Smith 1994, 1995). Corticospinal neurons are also sparse in the anterior parietal areas 3b and 1, but rise in areas 2 and 5. These projections are somatotopically organized, but whether they are functionally uniform is not clear.

There is a small contralateral corticospinal projection from SII in the macaque (Galea and Darian-Smith 1994; Burman 1992), and there are also a few corticospinal neurons distributed bilaterally in the caudal insular and retroinsular cortex (Burton 1986; Murray and Coulter 1981; Toyoshima and Sakai 1981).

2.4 Morphology of Corticospinal Neurons

2.4.1 Intracortical Organization of Corticospinal Neurons: Soma/Dendrite Morphology

Figure 2.10 illustrates the soma/dendritic form of identified corticospinal neurons in the pericentral cortex of the macaque, with those at the top and to the right located in motor area 4 and the remainder at the bottom and to the left in different somatosensory areas (3a, 3b). The plane of section was at right angles to the central sulcus. These and other labeled pyramidal corticospinal neurons in the same cortical areas (M. Sugitani, M.P. Galea, and I. Darian-Smith, unpublished) as well as cells located in parietal areas 2 and 5 and in the anterior cingulate cortex shared the following features:

- The soma was located within cortical lamina V
- An extensive apical dendrite which usually extended perpendicularly through to laminar I and most often consisted of a single shaft with many secondary branches
- A widely spreading skirt of basal dendrites located within laminae V and VI
- Many spines on most dendritic branches
- An axon arising from the soma or from the first-order segment of a basal dendrite, which commonly passed directly into the adjacent subcortical white matter and from which recurrent collaterals back into the cortex could often be identified

In examining a large sample of these intracellularly labeled neurons in the different corticospinal neuron populations in the mature macaque, the similarities of the dendritic morphology were more apparent than any regional differences. Of course, this qualitative assessment does not preclude regional differences in the geometry of the dendritic branching, but it does indicate that identifying these differences and assessing their functional relevance requires a systematic analysis similar to that of the rat's visual cortex reported by Larkman (1991a–c). Such analyses of identified corticospinal cells have yet to be completed in the primate cortex. The earlier, technically difficult in vivo labeling of such cells in area 4 of the macaque by Ghosh et al. (1988) demonstrated dendritic morphology visually similar to that illustrated in Fig. 2.10. However, the sample was insufficient to quantitatively assess differences in dendrite morphology.

In terms of providing insight into the circuitry of the cerebral cortex, in spite of their technical elegance, studies of the dendrite morphology of cortical pyramidal neurons have been disappointing. One reason for this past failure has been that usually the initial identification of the labeled neuron was incomplete. When this was achieved by antidromically firing the cell by electrical injection of fibers in the medullary pyramid (Ghosh et al. 1988), the neuron sample in monkeys was small. However, with the introduction of the injection of intracellular labels in fixed cortical tissue (Buhl and Lubke 1989), many corticospinal neurons can now

Fig. 2.10. Identified corticospinal neurons in area 4 (two cells at the *top* and the cell to the *right*), area 3a (two cells at the *bottom*), and a single cell in area 3b (cell on the *left*). These cells were first labeled retrogradely with fast blue (FB) by a prior injection into the contralateral dorsolateral column in the cervical spinal cord of the macaque. The brain was fixed and then cut into slices 200–400m thick. Single prelabeled cells were visualized using fluorescence microscopy and impaled with a micropipette. The impaled neuron was then injected iontophoretically with a mixture of lucifer yellow (LY) and biocytin (Buhl and Lubke 1989). Subsequently, a permanent label was produced with an immunochemical conversion of the injected fluorescent labels. A systematic analysis of regional variations in the morphology of these corticospinal neurons has yet to be completed. *x-axis*, mediolateral axis of cortex; *y-axis*, rostrocaudal axis of cortex; *z-axis*, corticospinal soma density (*DENS*) ▶

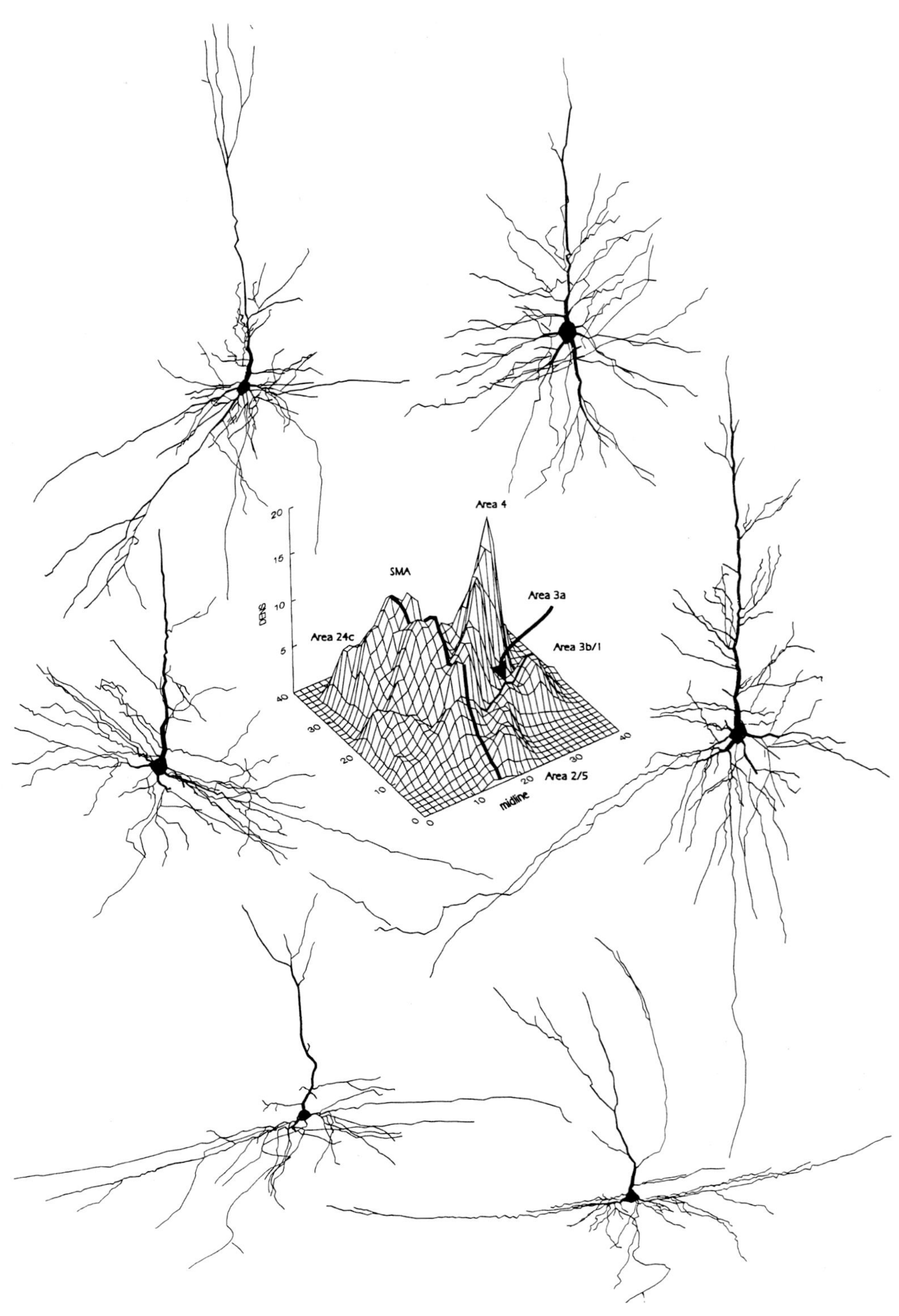

Area 4
SMA
Area 3a
Area 24c
Area 3b/1
DENS
Area 2/5
midline

be unequivocally identified. An essential, yet still insufficient step is to now develop an adequate *quantitative* description of the dendrite and axon morphology of the constituent neurons of each of the regional populations of these corticospinal cells. Methods of carrying out this analysis are now available. The next immediate and exciting challenge is to develop a quantitative description of the total synaptic input to individual corticospinal neurons, including the spatial distributions of the boutons of known origin and the statistics of this input to populations of these cells. By combining intracellular labeling of the postsynaptic neuron with anterograde labeling of specific presynaptic terminals, and identifying synaptic contacts using light and confocal microscopy with electron microscopy, these experiments are currently possible, but have yet to be done.

The dendritic/axonal morphology of corticothalamic neurons in the macaque (see Fig. 3.13), like those of the corticospinal cells in Fig. 2.10, are located in the pericentral cortex. The two populations are closely intermingled in the sensorimotor cortex. Qualitatively, the soma/dendrite pattern of many cells in these two distinctive populations of cortical neurons are similar, although the somas of most corticothalamaic neurons are located within lamina VI rather than lamina V. In addition, the apical dendrites of some corticothalamic neurons, unlike those of corticospinal neurons, do not extend perpendicularly through all cortical laminae (see Fig. 3.13). To gain insight into the different roles of these morphologically similar cortical neurons, we will require a much more quantitative account of the synaptic inputs to individual cells, and to populations of these cells, than we currently have.

2.4.2 Axon Terminations of Corticospinal Neurons

The use of multiple retrogradely and anterogradely transported tracers in individual monkeys has greatly increased the precision with which the origins and terminations of corticospinal neuron populations projecting to different target populations may be compared. This has encouraged a more global view of the different corticospinal projections and their interrelations, something that was difficult to achieve with the earlier "one-shot" axon terminal degeneration labeling, and HRP and autoradiographic tracer techniques, in which only one corticospinal population (two in a few studies) was mapped in each monkey. In the early studies (e.g., Leyton and Sherrington 1917, who traced degenerating fibers produced by a precentral cortical lesion in the chimpanzee, orangutan, and gorilla), projections were traced from the precentral cortex into the spinal ventral horn and were thought to terminate on motoneurons. Subsequently, it was recognized that only a fraction of corticospinal neurons have direct synaptic access to spinal motoneurons and that individual corticospinal cells in this group may synapse on motoneurons in one or several motoneuron pools (Fetz et al. 1976; Asanuma et al. 1979; Shinoda et al. 1979; Kuypers 1981; Lawrence et al. 1985). Fiber projections of the separate corticospinal neuron populations in areas 4, 6aα (including the supplementary motor area), and the anterior cingulate, insular, and parietal cortex each have terminals in distinctive areas of the spinal grey matter. This terminal distribution usually includes the intermediate zone, extending into the dorsal horn for projections from the parietal and insular cortex, and into the ventral horn for those from the frontal cortex (Kuypers 1981; Coulter and Jones 1977; Cheema et al. 1984; Ralston and Ralston 1985). Our present understanding of the spinal circuitry mediating complex movement, and of how the different corticospinal populations relate synaptically to this regional circuitry, remains poorly formulated. However, it does seem that the different interneuron populations that Eccles (1964) and others (Windhorst 1991; Lundberg 1982; Hultborn and Illert 1991) characterized as components in a series of stable spinal reflex pathways mediating well-defined, basic patterns of movement of the limbs may not operate as entities in the active, manipulating monkey. Rather, the coactive population of spinal neurons of which a particular neuron is a constituent may change with the motor task (see McCrea 1992).

Our current understanding of the structural organization of corticospinal axon terminals in the macaque may be summarized as follows:

- Each of the corticospinal projections from the cingulate, the SMA, precentral area 4, and the parietal and insular cortex extends to all segments of the spinal cord, and each has a distinctive distribution of axon branches with excitatory presynaptic terminals within these spinal segments (Fig. 2.11; Coulter and Jones 1977; Toyoshima and Sakai 1981; Ralston and Ralston 1985; Hutchins et al. 1988; Dum and Strick 1991; Bortoff and Strick 1993; Galea and Darian-Smith 1994, 1995).
- Each of these projections has some terminals in particular parts of the spinal intermediate zone (laminae V–VII), a region containing many intrinsic interneurons or propriospinal neurons (Fig. 2.11; Liu and Chambers 1964; Coulter and Jones 1977; Cheema et al. 1984; Ralston and Ralston 1985; Bortoff and Strick 1993; Armand et al. 1994).
- Some corticospinal neurons originating from precentral area 4 and the SMA, and possibly the cingulate cortex, but not parietal areas 3a, 3b, 1, or 2/5, also have terminals in the central and dorsolateral part of the spinal ventral horn (lamina IX), some of which synapse directly on motoneurons innervating the distal musculature of the limbs. Some corticospinal axons originating from the frontal and cingulate cortex also terminate in the neck of the dorsal horn (laminae IV and V), but not in the more dorsal laminae I and II (Ralston and Ralston 1985; Bortoff and Strick 1993).
- Some corticospinal neurons originating from the parietal (areas 3a, 3b, 1, 2, and 5) and insular cortex (SII) have localized distributions of terminals which extend from the intermediate zone into the medial neck region of the spinal dorsal horn (laminae IV–VI), but a few terminal branches also penetrate into the substantia gelatinosa and marginal zone (laminae I and II; Cheema et al. 1984; Ralston and Ralston 1985).
- Although there is an identifiable somatotopy within most of the corticospinal projections, this is not a point-to-point corticospinal projection. Rather, there is a *convergence* of input to a particular spinal segment from significant fractions of each of these corticospinal neuron populations. This spatial convergence is well shown in the contour maps of corticospinal projections from area 4 to cervical and lumbar spinal segments in Figs. 2.5–2.8 and 2.15 (Dum and Strick 1991; Galea and Darian-Smith 1994, 1995). Corticospinal input to the upper cervical and thoracic spinal segments is less dense and less con-

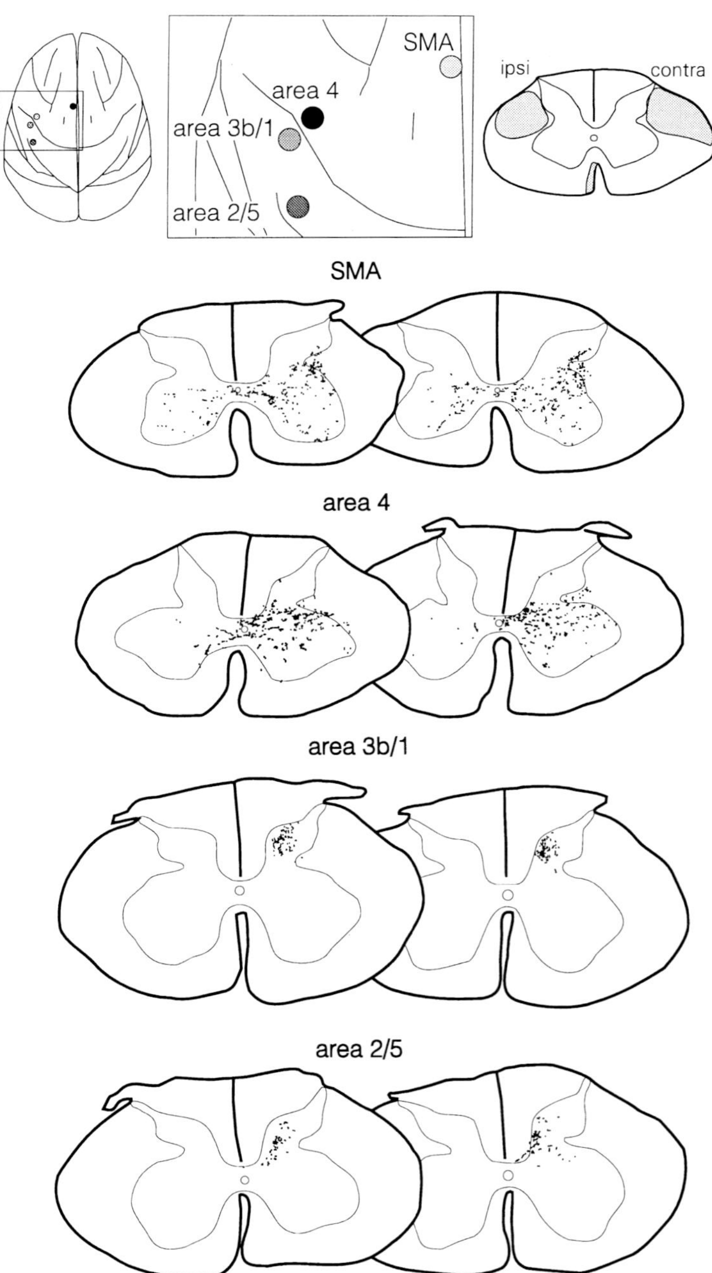

Fig. 2.11. Series of maps of axon terminals of corticospinal axons labeled anterogradely with biotinylated dextran amine which was previously injected into different areas of the sensorimotor cortex (supplementary motor area, *SMA*; areas 4, 3b/1, and 2/5) in the mature macaque. The pattern of terminal projection for each of these corticospinal neuron populations was unique. Corticospinal terminations of projections from the contralateral area 4 were concentrated in three zones; the lateral and medial parts of the intermediate zone and the more dorsal part of the ventral horn. There was a similar, but much less dense ipsilateral terminal projection. The projection from the SMA was even more widespread, on the contralateral side extending up into the neck of the dorsal horn and more ventrally into the ventral horn. The ipsilateral

vergent than to the cervical and lumbar enlargements. Figure 2.11 illustrates the convergence of axons originating from motor area 4 and the SMA to much of the intermediate zone and ventral horn in the macaque's cervical spinal cord; axons from both these cortical origins extend into the neck of the dorsal horn, along with others arising from parietal cortex (areas 3a, 3b, 1, 2, and 5). The synaptic organization of these multiple convergent projections is poorly understood.

- There is considerable *divergence* of the terminal axon branches of individual corticospinal neurons. Thus, the terminals of some single corticospinal neurons originating from area 4 may synapse not only on all the motoneurons that constitute the motoneuron pool of a particular muscle, but also synapse on motoneurons in a second or more synergistic motoneuron pool (Shinoda et al. 1981).

Mapping out the density of labeled corticospinal terminals (by terminal degeneration or with a particular anterograde tracer) in the different regions of the ventral horn of each spinal segment has been the method used for indirectly assessing the likelihood of corticomotoneuronal synaptic links or synaptic input to other neuron populations in the intermediate zone or in different elements of the dorsal horn (e.g., Leyton and Sherrington 1917; Coulter and Jones 1977; Tigges et al. 1979; Kuypers and Brinkman 1970; Kuypers 1981; Bortoff and Strick 1993). However, since this correlation ignores the dendritic tree (often more than 1 mm in diameter) as a postsynaptic target for corticospinal terminals, it is an uncertain procedure. Electron microscopy of identified pre- and postsynaptic elements is the only certain way of identifying synapses, but its use greatly limits the questions that can be answered. Examining the close apposition of labeled presynaptic terminals and intracellularly injected motoneurons with light or confocal microscopy can establish a reasonable probability of synaptic contact (e.g. Lawrence et al. 1985).

2.4.3 Corticomotoneurons and Manual Dexterity

Clearly, corticomotoneuron populations are likely to have an important role in shaping the use of the hand in the primate. Fortunately, important advances have been made in the systematic analysis of these neurons over the last decade. However, definition of the anatomy of corticospinal and regional spinal circuitry in the macaque is still fragmentary, and it is still poorly understood how these populations interact with each other to determine the monkey's clever use of its hands. Two main streams of experimental analysis have developed, which in

◀ projection was similarly widespread, but not as dense. The distributions of axon terminals of corticospinal projections from the parietal cortex (3b/1 and 2/5) were restricted to the medial half of the spinal dorsal horn, with some terminals extending into the marginal zone (lamina I) and substantia gelatinosa. Ipsilateral terminals of these projections were not present. The injection sites for the anterograde label (in different monkeys) are shown at the *top*, along with the sites of the descending fiber tracts. The duplicate maps illustrate the similarity of the distributions of boutons in adjacent transverse sections of the spinal cord

many respects complement each other, but differ in emphasis concerning some important issues. These are briefly examined below.

It has long been appreciated (Napier 1980) that the full repertoire of manual dexterity in the anthropoid primate depends on the capacity to move individual digits relatively independently and to oppose the thumb with another digit. In the 1960s and 1970s, the idea that the controlled, separate use of the digits depends on the integrity of the corticomotoneuronal populations was examined in several laboratories. Indirect evidence for this association was based on the following:

- The excitatory monosynaptic input of some corticospinal neurons projecting from cortical area 4 to motoneurons innervating hand muscles in the macaque (Bernhard et al. 1953) and baboon (Phillips and Porter 1964), but not in the cat or in other primates (Costello and Fragaszy 1988), such as the squirrel monkey, that lack the capacity to move the digits independently and to rotate the carpometacarpal joint
- The coincident acquisition in the infant macaque of independent finger movement and the invasion of the dorsolateral ventral horn by the terminal branches of corticospinal axons during the 6–8 months following birth (Kuypers 1962)
- The reported permanent loss of independent finger movement in the macaque following a *complete* bilateral pyramidotomy (Lawrence and Kuypers 1968a,b)

These studies have been largely, but not entirely, verified and refined by subsequent studies (e.g., Bortoff and Strick 1993; Flament et al. 1992a,b; Armand et al. 1994; Galea and Darian-Smith 1995). It should be noted, however, that all this evidence is indirect and incomplete.

In 1976, Fetz and coworkers (Fetz et al. 1976; Fetz 1980) developed a technique, spike-triggered averaging, for identifying corticomotoneuronal cells among the active corticospinal neuron populations in a macaque performing various manual tasks. This method depends on the expectation that, following monosynaptic excitation of a motoneuron by the input from a corticospinal neuron, the resulting excitatory postsynaptic potential (EPSP) facilitates the discharge of the motoneuron and of the muscle fibers it innervates. If this short-latency facilitation occurs, evident from close time-locking of the action potentials in the pre- and postsynaptic cells, then the probability is high that only one synapse is involved and that the activating neuron is a corticomotoneuronal cell. The special interest of these corticomotoneuronal populations lies in the fact that they directly affect motoneuron activity; there is a causal relation, and not simply a correlation (Fetz et al. 1989; Fetz 1992), between their activity and that of the motoneurons which they target. By using this procedure, some features of corticomotoneuronal cells in the alert, active monkey have been established, which are listed below:

1. Single corticomotoneuronal cells that are active during the execution of a precision grip innervate many low-threshold motoneurons in the motoneuron pool of at least one "hand" muscle and commonly of two or three

muscles of the nearly 30 (Lemon 1993) which regulate the macaque's hand movements.

2. The discharge pattern of individual corticomotoneuronal cells may vary greatly with the nature of the digital movement. Some of these cells are active during a controlled wrist movement, but silent during some ballistic movement of the wrist (Cheney and Fetz 1980). Similarly, some corticomotoneurons are more active during the execution of a precision grip than a power grip (Muir and Lemon 1983), although the muscles targeted by the cortical cell are active during each task.
3. The somas of corticomotoneurons that synapse on an individual spinal motoneuron may be somewhat irregularly spread out or clustered in lamina V of the motor cortex (area 4). Further, the soma distributions in lamina V of corticospinal neuron populations converging on single motoneurons in different motoneuron pools may overlap quite extensively.
4. The activity of a corticomotoneuron in area 4 during the execution of a particular manual task may differ in two ways relative to the activity in the *population* of these cells. First, when the actual population of active cells does not change, the discharge pattern of this cell may change with the task, and so its contribution to the activity of the cell population changes. Second, the active population of corticomotoneurons may change with the task, containing some new cells which were not previously active, and without input from other previously active cells. In each instance, the overall response of the active neuron population is a unique representation of different information.

As might be expected with such complex findings, different investigators have focused on somewhat different issues. Lemon and colleagues (Lemon 1993; Porter and Lemon 1993; Bennet and Lemon 1994; Maier et al. 1993; Buys et al. 1986; Lemon et al. 1986; Muir 1985) have examined in detail the role of corticomotoneuronal populations in the performance of different manual tasks, developing the ideas that Phillips and Porter outlined in the 1960s and 1970s (Phillips and Porter 1964, 1977). The corticomotoneuronal populations they have analyzed all originate from the precentral area 4, based on the view that none of the neurons in the remaining corticospinal neuron populations synapse monosynaptically on motoneurons in the cervical spinal cord. (However, see Fig. 2.11, which illustrates the terminal spinal projection of corticospinal neurons in the SMA in the macaque.) These investigators emphasize the unique importance of corticomotoneuronal connections in determining primate manual dexterity, almost to the exclusion of other neuron populations in the spinal cord. Lemon (1993) summarizes this view as follows:

> The development of corticomotoneuronal projections, in particular, allowed the motor hierarchy to bypass spinal segmental mechanisms, and to break up the rigid synergies of the spinal apparatus by direct access to the motoneurones, to the final common path itself.... The corticomotoneuronal system appears to play an important part in the fractionation of muscle activity allowing for independent movement of the digits under voluntary control.

Little consideration is given to the extensive corticospinal populations that terminate on the different local circuit neuron populations in the spinal cord.

A somewhat different emphasis has pervaded a second stream of analysis of the role of corticospinal neuron populations in manual dexterity, epitomized in the experimental studies of Fetz and colleagues (Fetz and Cheney 1980; Fetz et al. 1976, 1989; Cheney et al. 1988, 1991a,b). Fetz's views are thoughtfully summarized in his recent review (Fetz 1992). The central issue in this work is that the coding of information in corticospinal pathways about the movement patterns and the forces applied in the voluntary use of the fingers and hand is embedded in the responses of quite large *populations* of coactive cells, and not in the responses of single neurons. This is a tenet which, of course, applies to all neuronal transmission. Fetz emphasizes that sensorimotor information is transmitted from the cortex to the spinal motoneuron populations through the multiple *parallel* corticospinal pathways that we have described earlier, the corticomotoneuronal populations being one (or more) of these. Transmission of information about particular behavioral parameters of a manual task is in general *distributed* across these multiple coactive parallel pathways, rather than being relayed through a single neuron population. This does not mean that identical information is transmitted in each of these populations. Rather, there is specialization among these neuron populations, so that information about a particular behavioral parameter may be strongly represented in one or a few such projections, and only to a limited extent in others. This implies, as seen earlier, that the responses of single neurons may often provide little insight into the coding process of which they are part. Further, this parallel distributed processing implies that localization of function does not occur in the sense that particular corticospinal neuron populations are committed to the representation and processing of all the information about a particular parameter of manual behavior. Conversely, with this form of processing a particular corticospinal neuron population may contribute to the represention of several different features of manipulative behavior. At the systemic level, parallel distributed processing implies that even with the simplest manipulative tasks several corticospinal neuron populations will be coactive; we have seen that this is so. At a more microscopic level of organization, the averaged response of the neuron population may be necessary to represent key information. Thus the coding by motor units of the force developed by the muscle they innervate is in terms of this summed population response, because the dynamic range of each constituent motor unit spans only a small part of the total force developed by the muscle; this is parallel distributed processing at a cellular rather than systems level. Thus, parallel distributed processing seems to be important at both these levels of organization.

Corticomotoneuron populations constitute an important element of all corticospinal populations, because they have an experimentally identifiable *causal*, and not just correlative, relation to the patterns of voluntary hand movement that are generated (Fetz 1992). However, in any realistic model of corticospinal action in manipulative behavior it is essential to consider the roles of *all* the corticospinal neuron populations involved, as well as the coactive populations of spinal local circuit neurons and, of course, the relevant motoneuron populations.

The characterization of corticomotoneurons considered above has been of great value, both in elucidating their functions and in highlighting some of the limitations of current knowledge and the direction for further investigation. For example, the systematic analysis of manual behavior in the adult macaque, in the developing infant, and following localized lesions in the corticospinal pathways needs to be substantially extended to obtain careful quantitative accounts of digital, hand, wrist, and forearm movements and of the identifiable corticospinal connections characterizing each experimental monkey used. We are far too dependent on the elegant and seminal, but narrowly focused studies of Lawrence and Kuypers (1968a,b; Kuypers 1981) on the effects of pyramidotomy on hand usage. The anatomical verification of corticomotoneurons needs to be rigorously developed; currently, too much evidence advanced for identifying these cells depends on the spatial overlap of visualized corticospinal teminals and the distributions of spinal motoneuron somas.

2.5 Indirect Corticospinal Projections

In parallel with the direct corticospinal projections, there are several less direct corticospinal connections, which include corticorubrospinal (for a review, see Keifer and Houk 1994), corticoreticulospinal (Grantyn et al. 1993), and corticovestibulospinal projections (Akbarian et al. 1992). We consider these complex pathways quite selectively, comparing their structural organization with that of the direct corticospinal pathways and examining what the similarities and differences may imply. A bulbospinal neuron population is the core component of each of these pathways; each has a major noncortical input and a smaller, often circuitous cortical input. Figure 2.12 illustrates this complex organization for the corticorubrospinal connection in the macaque. The operation of each of these pathways and the contribution they make to voluntary, adaptive motor behavior are poorly understood, but the fact that they each contribute to the maintenance of patterns of sustained contraction in the limb and trunk muscles of mammals has been appreciated since the decerebrate cat and monkey were studied in detail by Sherrington (1898, 1947) and others. As seen earlier, Kuypers (1981) grouped these bulbospinal and the direct corticospinal descending pathways on the basis of their topography and patterns of termination in the spinal cord. Both the corticospinal and rubrospinal somatotopically organized pathways in the macaque are predominantly contralateral and descend in the spinal dorsolateral column, and their terminal branches substantially overlap in all spinal segments in the dorsal and lateral parts of the intermediate zone (see below). Functionally, the neuron populations in these two pathways are likely to be complementary, but how their action is coordinated is not known and has triggered some speculation (Kennedy 1990). Kuypers' (1981) second group of descending spinal pathways includes the vestibulospinal and reticulospinal neuron populations; these mainly bilateral pathways are located in the ventral and ventrolateral spinal white columns and terminate in the ventromedial part of the intermediate zone of spinal gray matter. The reticulospinal neuron populations in the mammal are distributed within the midbrain, pons, and medulla oblongata, and not all (e.g. the spinal projection from the raphe nuclei, concerned with the modulation of nociceptive afferent input) relate to the control of voluntary movement. Even those reticulospinal pathways thought to have a sensorimotor role are heterogeneous. Projections from the ventrolateral pons descend mainly in the contra-

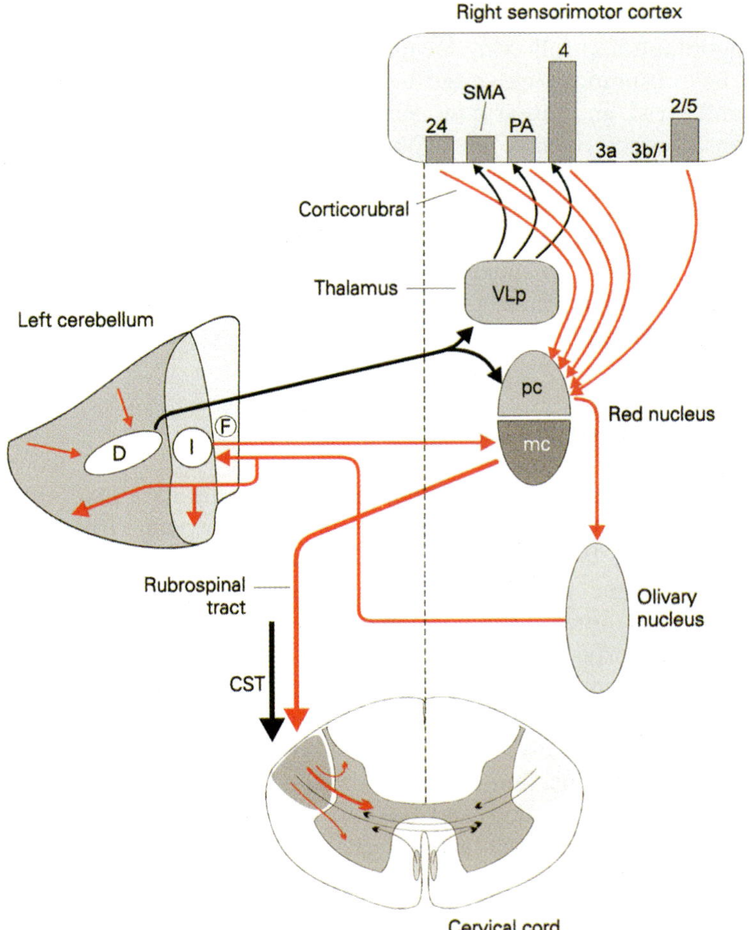

Fig. 2.12. Corticorubrospinal connections in the macaque. Little cross talk occurs between the two components of the red nucleus, the parvicellular (*pc*) and magnocellular (*mc*) nuclei. Most corticorubral fibers, originating from the frontoparietal cortex, terminate in the parvicellular nucleus. The somas of most rubrospinal neurons are within the magnocellular nucleus, and their axons cross the midline and descend with the direct corticospinal fibers in the spinal dorsolateral column. The main output of the parvicellular nucleus is to the inferior olivary nucleus, from which the climbing fiber input to the cerebellum originates. Pathways in *red* indicate a possible, somewhat roundabout corticospinal channel. *D*, dentate nucleus; *I*, interpositus nucleus; *F*, fastigial nucleus; *SMA*, supplementary motor area; *VLp*, ventral lateral posterior nucleus; *CST*, corticospinal tract; *Pa*, postarcuate cortex

lateral dorsolateral column along with the corticospinal and rubrospinal projections. However, the bilateral projections from the medial reticular nuclei in the midbrain and medulla descend in association with the vestibulospinal axons in the ventral and ventrolateral white columns of the spinal cord.

Here, only the corticorubrospinal pathways are considered, and they are used to illustrate some of the features of these indirect projections. The reader is referred to recent analyses of the corticovestibulospinal (Akbarian et al. 1992,

1994; Fredrickson and Rubin 1986) and corticoreticulospinal (Grantyn et al. 1993) connections.

2.5.1 Corticorubrospinal Projections

Most flowcharts, like that shown in Fig. 2.12 of the connections of the red nucleus in the macaque, are selective and based on a model of the likely sequential transfer of information. Figure 2.12 attempts to provide some insight into two questions:

1. What connections, if any, would permit the transfer of sensorimotor information from the cerebral cortex to the motoneuron populations of the cervical spinal cord in the macaque?
2. What kind of neuronal processing might occur in the course of this transmission?

The red arrows identify the successive neuron populations that could constitute a substantial corticorubrospinal pathway. The functional anatomy of this pathway is considered before examining the sort of processing that could result from transmission along it.

The red nucleus in the macaque and other primates consists of two spatially separate components, which appear to have no direct interconnections; both the larger rostral parvicellular nucleus (Rpc) and the caudal magnocellular nucleus (Rmc) are packed with medium-sized and small neurons, and the Rmc additionally contains neurons of quite large diameter from which the rubrospinal axons originate (Keifer and Houk 1994; Nathan and Smith 1982). The small cells in each nucleus are GABAergic interneurons (Ralston and Milroy 1992; Ralston and Ralston 1993, 1994). The rubrospinal neuron somas are somatotopically arranged; those projecting to the cervical spinal cord are located in the dorsomedial Rmc and those projecting to the lumbar cord are in the ventrolateral part of the nucleus (Holstege et al. 1988; K. Burman, M.P. Galea, and I. Darian-Smith, unpublished). The crossed rubrospinal axons descend in the dorsolateral column of the spinal cord, intermingled with corticospinal axons, and they terminate mainly on interneurons in the dorsolateral intermediate zone of the gray matter of up to three spinal segments (in the cat; Shinoda et al. 1988). Some axons, however, do synapse directly on dendrites of motoneurons in the ventral horn (Ralston et al. 1988). Few if any rubrospinal neurons are within the Rpc, whose major projection is to the olivary nucleus (Fig. 2.12).

The major inputs to the red nucleus are from the contralateral deep cerebellar nuclei (Fig. 2.12), the input to the Rpc coming from the dentate nucleus (D) and that to the Rmc originating from the nucleus interpositus (I). The cerebellar excitatory terminals in both subnuclei contact the somas and proximal dendrites of rubrospinal neurons, but do not seem to synapse on the interneurons (Ralston 1994a).

About 90% of corticorubral axons terminate in the Rpc (Humphrey et al. 1984), with a dense projection from the more rostral part of area 4 and less dense projections from area 6aα (from the SMA and the postarcuate cortex), from the caudal cingulate cortex, and from parietal area 5. This contrasts with the origins of the more extensive corticospinal projections, which include the above cortical areas and the rostral cingulate cortex, somatosensory areas 3a, 3b, 1, and 2, and the insular cortex (Galea and Darian-Smith 1994). The somas of both corticorubral and corticospinal neurons are all located within lamina V of the different areas of cortex in which they are located.

The remaining neuron populations which constitute the corticorubrospinal path highlighted in Fig. 2.12 are the olivocerebellar projections and the neuron populations of the deep cerebellar nuclei and of the intermediate and lateral zones of the cerebellar cortex. Olivocerebellar neurons originating from different subnuclei in the inferior olivary complex are unique in that they are the sole populations of cells with climbing fiber terminations on Purkinje cells in the whole cerebellar cortex. Since they can alter the response characteristics of Purkinje cells for a long period, they have been the focus of various theories of motor learning (Marr 1969; Ito 1984; Thach et al. 1992). The cerebellar projection from the nucleus interpositus to the Rmc is the final link in the complex corticorubrospinal pathway.

Single neuron recording in the Rpc, the Rmc, and the interpositus nucleus (Thach 1978; Kennedy et al. 1986; Mewes and Cheney 1991, 1994) reinforce the ideas suggested by the functional anatomy of these connections and imply that information transmitted by rubrospinal neuron populations is determined largely by the cerebellum, since the responses of neurons in the interpositus nucleus and the Rmc recorded in a monkey during the execution of various manual tasks (Keifer and Houk 1994) are quite similar and signal information about the various phasic components of the movement. However, contrasting with this, neurons in the Rpc lack this phasic activity during the execution of the same manual tasks. In other words, the direct cortical input to the red nucleus (Rpc) does not signal information about the intended or actual movement pattern, but rather about more general behavioral features, such as the initiation of the movement or the strategy of its execution. The involvement of the neuron populations in the inferior olivary complex in the indirect cortical projection to the Rmc suggests that corticorubrospinal pathways in the macaque may be implicated in motor learning.

The omission of some identified connections in Fig. 2.12 was necessary to avoid confusion. The small cortical input to the Rmc has been mentioned. In the cat, somatosensory inputs from the cuneate nucleus (Boivie 1988) as a direct spinorubral projection (Padel et al. 1986; Padel and Relova 1988) have been identified, as have cutaneous receptive fields of single neurons in the red nucleus. Rubrospinal axons have collateral terminals in the contralateral brain stem reticular nuclei, which in turn send mossy fibers to the cerebellar cortex.

2.6 Spinal Circuitry Mediating Supraspinal Control of Movement

After a century of intensive analysis, the morphology and response characteristics of *individual* neurons from the different populations in the spinal cord are known perhaps better than any other neuron types in the primate brain. This is certainly true for the motoneuron, for intrinsic neurons in the dorsal horn, and for those spinal neurons which project to the brain stem, cerebellum, and thalamus. Characterization of single neurons in the various intrinsic spinal populations has been less complete, but still extensive. Nevertheless, in spite of this accumulation of data, we still do not have a satisfactory picture of how these multiple cell populations in the spinal cord operate together to account for any voluntary movement and which specifically mediate the intelligent manipulation of objects within reach. Even accepting a severely reductionist approach, no satisfactory model of the spinal circuitry that mediates complex movement initiated from the forebrain has yet emerged (Hultborn and Illert 1991).

Although a convincing global picture of sensorimotor processing in the spinal cord has yet to be developed, two streams of research, each with a long history, have focused on relating particular spinal neuron connections, somewhat to the exclusion of others, to this action. The first stream, initiated by Sherrington, and later greatly developed by Eccles (1964) and Lundberg (1966), placed great emphasis on the different spinal interneuron populations interposed between the various inputs to each spinal segment and its motoneuron outputs, which subserved different types of stereotyped "reflex" activation of the motoneuron. Complex movement known to be controlled by the forebrain was viewed as resulting from the supraspinal modulation of the basic spinal reflex activity and was thought to be achieved mainly by regulating the activity of particular interneuron populations in the various reflex pathways. These pathways were considered to be sequential and essentially an elaborate *relay* system, which ultimately played the right tune on the appropriate motoneuron populations. Feedback from the periphery was recognized to have an important role in shaping the resulting pattern of movement. The second stream of investigation, promoted by Kuypers (1981), emphasized the importance of supraspinal synaptic input to spinal motoneurons in the acquisition and maintenance of fine manipulative action involving the independent movement of digits. With this model, again a relay, the role of spinal interneuron populations becomes uncertain, but presumably they can modulate transmission in the corticomotoneuronal pathways. Neither approach has resulted in a balanced, inclusive view of spinal processing, but both have generated valuable new information about connections within the spinal cord, which must ultimately be incorporated into any useful model of corticospinal action. The extensive convergent segmental, intersegmental, and supraspinal inputs to the spinal interneuron populations, along with the divergence within each of the corticospinal projections, implies that spinal intrinsic neuron populations (interneurons and propriospinal neurons) do *not* operate serially, but constitute elements in multiple local circuits which operate in parallel. Further, information relevant to all common patterns of movement is likely to be represented in the simultaneous activity in different composites of the various local circuit neuron populations; in other words, parallel distributed processing, rather than serial processing, is the norm. Recognition of this framework for neuronal processing in the spinal cord is unfortunately only the first step in understanding the role of the spinal cord in voluntary action. A much more rigorous description than we currently have of spinal circuitry and its operations is needed before its contribution to voluntary action can be understood.

Fortunately, in recent years the attractive anatomical techniques of axon tracing and the intracellular injection of label to visualize the morphology of functionally or anatomically identified neurons have encouraged investigators in both research streams to focus on the structural organization of spinal neuronal connectivity, and to this extent their analyses are complementary.

2.6.1 Functionally Defined Interneuron Populations: Segmental and Propriospinal Connections

Major reflex circuits, or "spinal functional units" (Baldiserra et al. 1981), which have been separated out from the total spinal neuron populations, and those whose associated interneuron populations have been carefully studied include the following:

- The recurrent inhibitory pathway in which α-motoneuron recurrent collaterals activate Renshaw interneurons, which in turn inhibit motoneurons innervating the same or synergistic muscles
- Pathways activated by muscle spindle afferents, which include spinal interneurons which inhibit the action of α-motoneurons supplying antagonistic muscles
- Pathways activated by Golgi tendon organ afferents which include interneurons that inhibit α-motoneurons supplying the same muscles
- Multisynaptic pathways activated by groups II and III, designated by Eccles and Lundberg "flexor reflex afferents," and with input from group II muscle afferents and cutaneous afferents

From these studies, mainly in the cat, the structural characteristics of the various populations of spinal interneurons found in the different "reflex pathways" may be briefly stated as follows:

- There is diversity in the morphology of cells in the different interneuron categories, not only among those in the different reflex pathways, but also those within the simplest circuits. Thus two groups of morphologically distinct Renshaw cells have been described (Fyffe 1990), and several groups have been identified within the flexor reflex afferent (FRA) pathways (for reviews, see Jankowska and Edgley 1993; McCrea 1992).
- There is great convergence of input to most interneurons, from primary somatosensory neurons, from other interneurons, from propriospinal neurons, and from the various descending projections, including corticospinal neurons. However, this pattern of convergent input varies considerably for individual interneurons. Monosynaptic access of these spinal afferents to motoneurons does occur, but probably constitutes only a small fraction of afferent terminals.
- The neurons targeted by individual interneurons are diverse and extensive; divergence is the rule. Some, but not all interneurons have direct synaptic access to motoneurons.

It is ironic that, with all the previous magificent experimental analysis of spinal cord organization, once again its circuitry and operation become a central issue in understanding the neuronal processes mediating voluntary action, and specifically those underlying manual dexterity. In recent years, the persisting concept of reflex spinal pathways has not helped in this analysis. However, the multiple new anatomical methods for analyzing neuronal circuitry should be helpful in the immediate future. A second factor limiting progress in the analysis of spinal function has been the technical difficulty of recording from single neurons in the cervical and lumbar cord in active, alert experimental animals. If progress is to be made with spinal processes mediating hand function, then this recording must be done in macaques or other primates.

2.7 Postnatal Maturation of Corticospinal Pathways

2.7.1 Maturation of Manual Dexterity in the Macaque

Prehension and the handling of objects within reach are essential for the survival of the newborn macaque, and so, as might be expected, the maturation of manual behavior is quite rapid. During the first 2 months after birth, the monkey uses its hands in climbing, exploring the immediate environment, and grasping food, but is hesitant, clumsy, and slow. During the next few months, the infant fingers objects more quickly and precisely, and by about 8 months it has developed a manual dexterity approximating that of the mature macaque. Early in this review (Fig. 1.1), the trajectories of the finger and thumb that are executed when a macaque reaches for, grasps, and retrieves a small object were illustrated diagrammatically. Comparing the performance of this manual task in the 3-month-old infant and the mature monkey highlighted that (a) each phase of the movement pattern is slower and more dependent on sensory feedback in the infant than in the adult monkey and (b) both visual and tactile feedback of information concerning the position of the digits and the target object is especially important in the infant. Figure 2.13 further illustrates the rapid postnatal transition in the monkey's ability to retrieve a small object, opposing the index finger and thumb to do so. Two months after birth, the infant took approximately 2.9 sec to retrieve the object, whereas the 8-month old monkey completed the task in approximately 1.1 sec.

2.7.2 Maturation of Direct Corticospinal Projections in the Rodent and the Macaque

Does the acquisition of this dexterity reflect postnatal changes in the structural and functional organization of the corticospinal connections, or is this motor development entirely cognitive? More than 30 years ago, Kuypers (1962) showed in the macaque that at birth few corticospinal fibers have yet invaded the ventral horn and that the maturation of these axon terminals in close association with motoneurons, possibly with the formation of corticomotoneuronal synapses, is not completed until the sixth to eighth month. More recent studies (Felix and Wiesendanger 1971; Flament et al. 1992) using focal electrical stimulation of the motor cortex, area 4, have demonstrated a functional parallel; in the first 2–3 postnatal months, electrical stimulation of the area of hand representation in

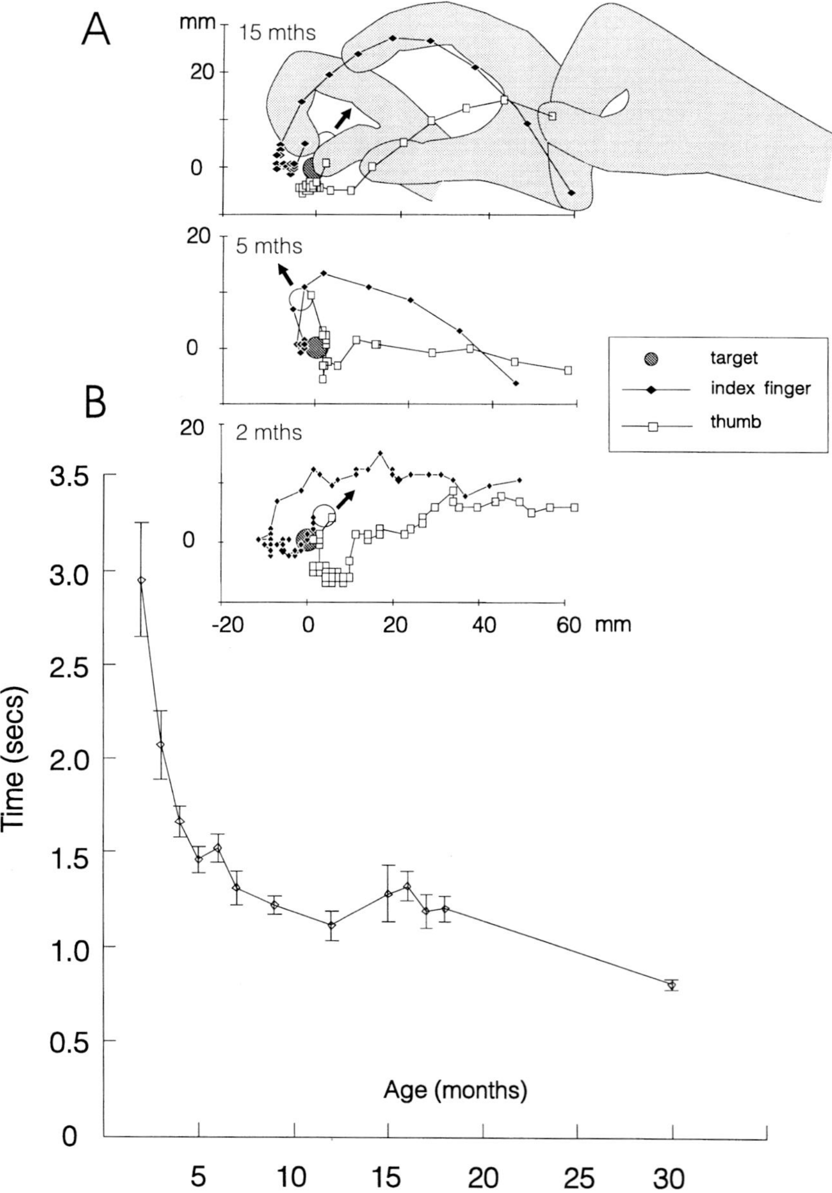

Fig. 2.13A,B. Maturation of finger/thumb opposition in the infant macaque monkey. The task, which required no training, was for the monkey to reach out through a small window to grasp a cylindrical glucose target (4 × 5 mm) held by a clip; the force needed to release the object was approximately 1.0 N. The finger and thumb had to be aligned vertically when gripping the target. **A** Trajectory of pads of thumb and index finger in the same monkey at 2, 5, and 15 months. The time interval between successive data points was 40 ms; the trajectory was measured over 60 mm; the stationary circular target is *shaded* and when displaced is *white*. **B** Time taken to reach out and displace target object in monkeys aged 2–30 months. Standard error bars are plotted. The 2-month-old monkey took about three times longer to execute the task than the mature animal, but by the eighth month was quite adept (Galea and Darian-Smith 1995)

motor area 4 fails to elicit movements, but by the fourth month short-latency finger or hand movements can obtained, and 8 months after birth the fast, brisk finger movements that typify the mature monkey can be obtained using low-intensity current. These studies together suggest that substantial postnatal changes may occur in the structural organization of the direct and indirect corticospinal neuron populations, which are not limited to the terminal branching in the spinal cord of axons projecting from motor area 4. The following questions arise: (a) are spinal projections from all the corticospinal neuron populations in the cingulate, frontal, parietal, and insular cortex slow in developing their mature pattern of terminations at each level in the spinal cord of the primate? (b) What postnatal changes occur in the different areas of cerebral cortex from which the different corticospinal projections originate?

Extensive studies of the developing corticospinal projections in the newborn rodent have been reported in recent years (Schreyer and Jones 1982, 1988a,b; Reh and Kalil 1981; Kalil 1984; Stanfield et al. 1982; Stanfield and O'Leary 1985; O'Leary and Stanfield 1986; Heffner et al. 1990; Stanfield 1992). Although these studies have differed somewhat in detail, the common finding has been that corticospinal projections in the newborn rat originate from the whole (including visual cortex) or a large area of the neocortex and that this exuberant projection regresses in the weeks following birth to assume the mature localized projection which originates mainly from the frontal cortex. Schreyer and Jones (1982, 1988a,b) also found a changing rostrocaudal level of termination of corticospinal axons in the rat's spinal cord. Axons which have grown into the lumbar spinal cord by the end of the first postnatal week are permanent and do not regress, whereas many of the corticospinal axons whose terminals have only reached the cervical cord at the time of birth subsequently regress. While the postnatal maturation of corticospinal neuron populations in rats proved to differ greatly from what was subsequently observed in the primate, its study did suggest that major postnatal changes in the macaque's corticospinal neuron populations are likely and that these occur both within the cerebral cortex and within the spinal cord.

2.7.2.1 Cortical Origins

Figure 2.14 illustrates a series of coronal maps of the distribution of corticospinal neuron somas that were retrogradely labeled following the injection of FB into the left dorsolateral column of C5 (see also Fig. 2.15). This monkey was delivered about 2 weeks prematurely by cesarean section on embryonic day 158 ± 3 (Galea and Darian-Smith 1995), and the labeling dye was injected 2 days later. All the labeled somas were in cortical lamina V. Comparing this contour map of the distribution of corticospinal neuron somas in the premature infant macaque with the matching map in a mature monkey (e.g., the top map in Fig. 2.8) highlights the fact that the separate concentrations of corticospinal neurons in areas 4, 6aα (SMA and postarcuate cortex), the cingulate, area 2/5, and the insular cortex, which characterize the distribution in the mature macaque, are also readily identifiable in the newborn monkey. There are, however, two important differences in the distributions of these somas in the newborn and mature macaque. First, even though at birth the brain is approximately 30% smaller than the

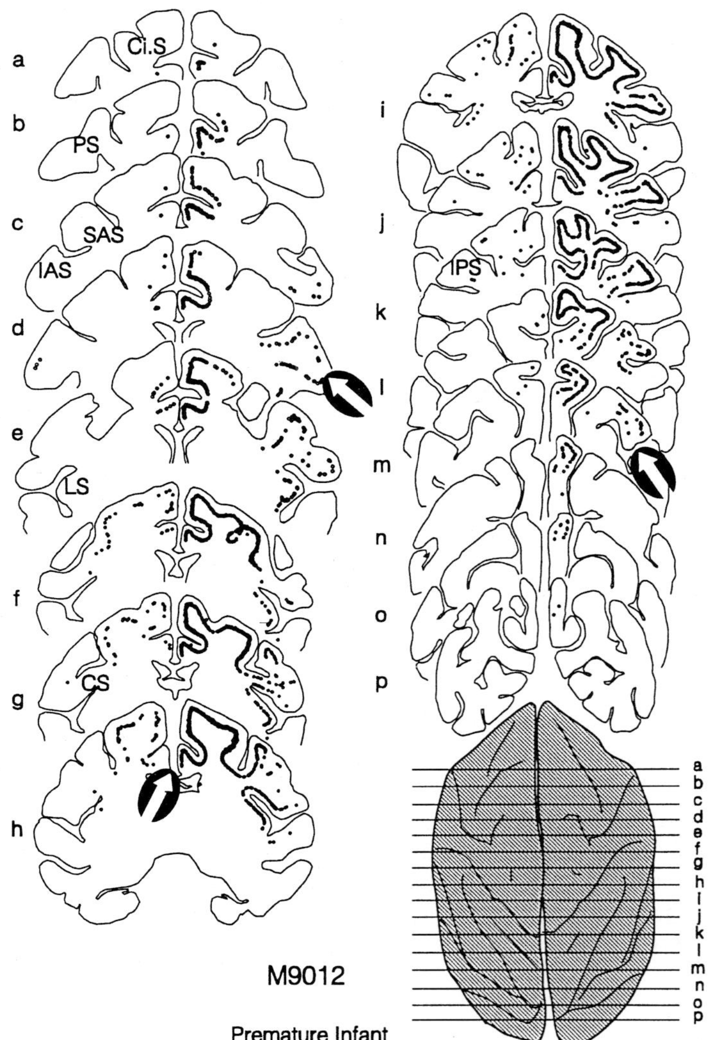

Fig. 2.14a–p. Coronal maps of the distribution of somas of labeled corticospinal neurons following the injection of a retrogradely transported label (fast blue, FB) in the left cervical cord at C5. This macaque was prematurely delivered by cesarian section, and label was injected on gestation day 16 ± 3 days, i.e., about 2 weeks prematurely. All labeled somas were within lamina V of the cortex. *Arrows* point to areas from which corticospinal projections originate in the newborn, but not in the mature monkey. *Ci.S*, cingulate sulcus; *PS*, principal sulcus; *SAS*, superior arcuate sulcus; *IAS*, inferior arcuate sulcus; *LS*, lateral sulcus; *CS*, central sulcus; *IPS*, intraparietal sulcus (Galea and Darian-Smith 1995)

mature brain, the areal extent of the corticospinal soma distribution in the infant is greater than in the mature monkey. In addition, the local soma densities are greater in the infant, as is apparent from the corresponding peaks in the two contour maps (see Fig. 2.15 and top map in Fig. 2.8). Galea and Darian-Smith (1995) have systematically examined these changes in the macaque during the first 2 postnatal years. Both the areal extent of the cortical origin (i.e the

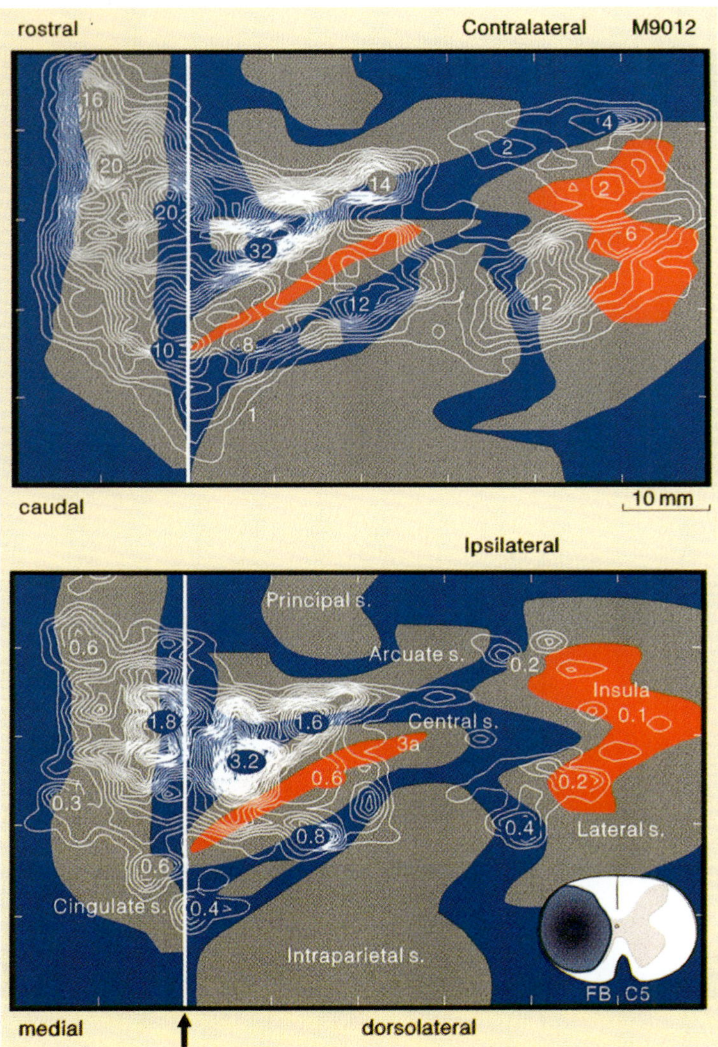

Fig. 2.15. Contour maps of soma distributions of contralateral and ipsilateral corticospinal projections to spinal C5 and more caudally, in the premature infant macaque described in the legend of Fig. 2.14. The map of left hemisphere is plotted as a mirror image to simplify the comparison of the maps. Lamina V of unfolded sulci is *gray*, and of area 3a and and insular cortex is *red*. The junction of the mesial and dorsolateral cortex is indicated by the *heavy white line*. Although the ipsilateral corticospinal projection is less than 10% of that from the contralateral cortex, both are characterized by separate neuron populations located in motor area 4, anterior cingulate area 24, mesial area 6aα (supplementary motor area, SMA), and postarcuate, posterior parietal, and peri-insular cortex. *s.*, sulcus; *FB*, fast blue (Galea and Darian-Smith 1995)

corticospinal soma distribution) and the relative numbers of corticospinal neurons with axons projecting to or caudal to the cervical cord, were measured (Fig. 2.16). Both parameters of the corticospinal neuron populations regress very substantially over this period; the reduction in the cortical area from which the

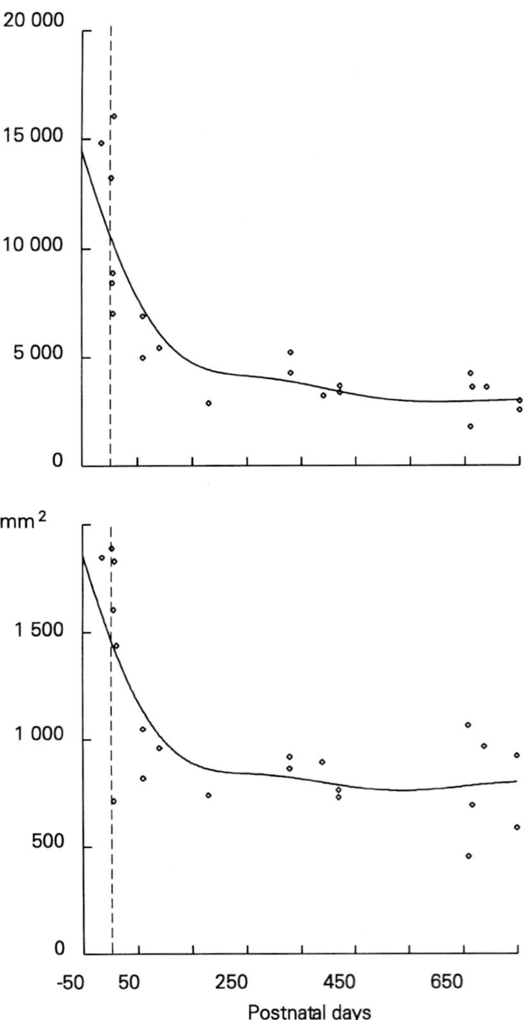

Fig. 2.16. Changes in the numbers (corticospinal tract, CST, count, *top*) of corticospinal neurons and the areal extent (CST area, *bottom*) of the distribution of their somas in macaques during the first 2 postnatal years. These estimates are made from the same data base that is used to construct the contour maps and three-dimensional representations of corticospinal soma distributions which project to the cervical spinal cord and more caudally (Galea and Darian-Smith 1995). The corticospinal neuron count fell threefold in the first 6–8 months following birth, and the areal extent of the distribution halved over this same period, even though the total area of neocortex increased by approximately 30%. Regression lines have been fitted to the data obtained from 21 monkeys (Galea and Darian-Smith 1995)

corticospinal fibers originated was approximately twofold, and the reduction in the count of labeled corticospinal neurons was even greater (i.e., threefold). Figure 2.16 illustrates that most of this considerable postnatal regression in the corticospinal neuron population occurred during the first 150 days after birth, the very period during which the infant macaque becomes manually dextrous.

2.7.2.2 *Spinal Terminations*

The large transient populations of corticospinal neurons in the newborn macaque can be visualized by retrogradely labeling the axons descending in the dorsolateral column in the cervical spinal cord. By contrast, relatively few of these corticospinal axons in the infant terminate in the spinal gray matter, and these arborizations are confined mainly to the spinal intermediate zone, with few terminals extending into the ventral and dorsal horns. Kuypers (1962) showed that corticospinal axon terminals invade the ventral and dorsal horn in the months following birth, and only by the eighth postnatal month is the terminal arborization comparable with that seen in the mature macaque. Both the exuberant corticospinal axonal projection and the sparse terminal arborization of these fibers can be visualized in the same animal by using several differentiable retrogradely transported tracers. In Fig. 2.17, three-dimensional maps of the soma distributions (all contralateral) in newborn and mature macaques are compared. The maps of soma distributions in the newborn monkey illustrate the exuberant axonal (dorsolateral column) and sparse intrasegmental terminal branching. In the mature macaque (two lower maps in Fig. 2.17), the total corticospinal projection has regressed, and the intrasegmental terminal branches of these same axons have invaded the dorsal and ventral horns. We have not observed a rostrocaudal gradient of maturation in the macaque's spinal cord. At birth, corticospinal fibers have already reached lumbar spinal segments, and the somatotopy of their cortical origins is apparent. The invasion of the dorsal and ventral horns by corticospinal terminals seems to occur at about the same time in the cervical and lumbar spinal segments.

The major postnatal structural changes in the different direct corticospinal neuron populations described in the preceding paragraphs and the establishment of corticomotoneuronal synapses (Flament et al. 1992a,b; Lemon 1993) occur over the same period of 6–8 months in which the infant macaque develops a moderate manual dexterity (see Armand et al. 1994). Just how closely these neuronal changes determine the behavioral maturation remains uncertain as long as the spinal circuitry involved and the postnatal synaptogenesis both in the forebrain and spinal cord are poorly specified. Nonetheless, Lemon (1993) and others (Porter and Lemon 1993) argue that the development of manual dexterity is directly linked to the coincident establishment of corticomotoneuronal connections.

While the postnatal shaping of the circuitry mediating manual dexterity and other complex sensorimotor behavior has an intuitive attraction, the role of an overproduction of a population of neurons, many of whose axons never enter the target zone, is less obvious. Recent experiments (Galea and Darian-Smith 1995), suggest that the postnatal regression of corticospinal neurons results in part from the selective elimination of axon collaterals, although neuronal cell death may also occur. The axons of those cortical neurons in the fetus which will establish synaptic input to neurons in the cervical spinal cord must traverse up to 10 cm to reach their target neuron populations and then develop synaptic connections with specific spinal cells. These two processes may be mediated by different, if not independent neuronal mechanisms (Stanfield 1992; Kennedy and Dehay 1993; O'Leary et al. 1994; Allendoerffer and Shatz 1994). The "pathfinder" process

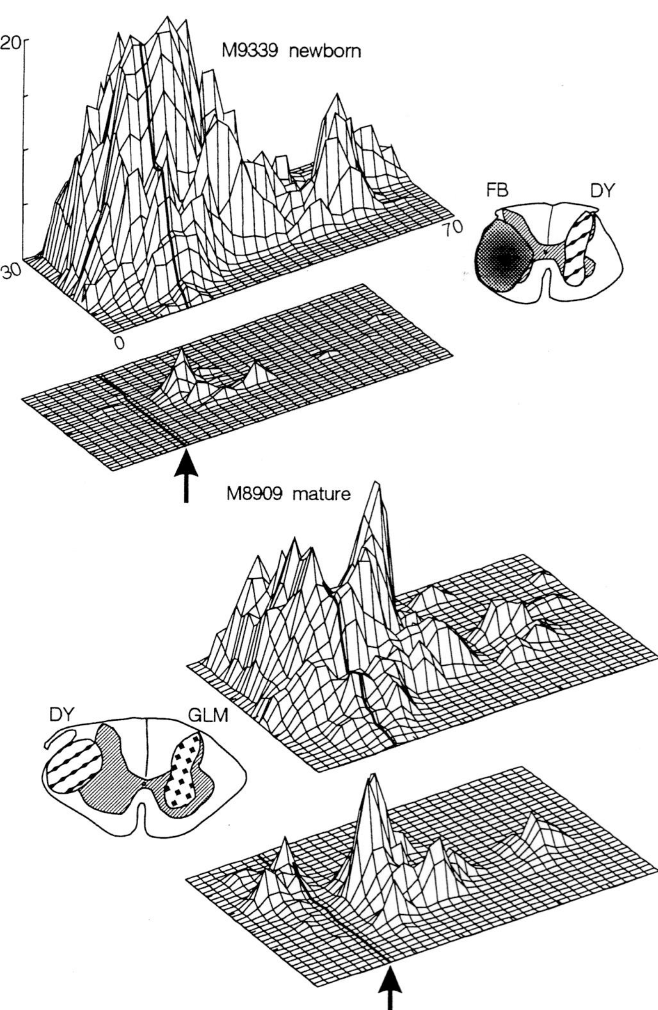

Fig. 2.17. Three-dimensional maps of the soma distributions of corticospinal neuron populations in newborn (*upper two maps*) and mature (*lower two maps*) macaques. The injection sites of the retrograde labels in the contralateral cervical spinal cord (C5–7) are shown in the *insets*. The upper map of each pair illustrates the soma distribution of corticospinal neurons whose axons traverse the cervical spinal cord in the dorsolateral column, i.e., all corticospinal fibers which terminate caudal to C5–7. The lower map of each pair illustrates the soma distribution of corticospinal neurons whose axons terminate within the contralateral spinal segments C5–7. In the newborn macaque, the total corticospinal neuron population is substantially larger than in the mature monkey. However, in the newborn few corticospinal axon terminals have yet invaded the gray matter of the contralateral cervical cord relative to those which terminate in this same zone in the normal mature monkey (Galea and Darian-Smith 1995). To simplify the comparison of the soma distributions, each map has been plotted onto the right hemisphere. *FB*, fast blue; *DY*, diamidino yellow; *GLM*, green latex microspheres

enabling the axonal growth cone to reach the proximity of the targeted neuron population has been modeled in terms of a cascade of specific local guidance cues (Dodd and Jessell 1988; Stanfield 1992). It has been proposed (for a review, see Allendoerfer and Shatz 1994) that a transient neocortical structure, the subcortical plate, provides a framework of mechanical and chemical cues to guide the axons emerging from neurons in laminae V and VI into the internal capsule. Neurons in the subplate develop transient axonal projections to different zones of thalamic space and to the tectum, and these may act as pioneer fibers to guide the permanent cortical projections to their respective target areas. Since nearly all cortical projections to the brain stem and cord also traverse the internal capsule, the cortical subplate may also have a role in funneling most cortical efferents originating from the deep cortical laminae into the internal capsule. The large excess of corticospinal axons confined within the internal capsule possibly provides the mechanical scaffolding for those axons which will survive and reach their bulbar and spinal targets. Relatively few mechanical and chemotropic cues may then be needed to guide the descending corticospinal axons into the regions that are the forerunners of the pyramids and white columns of the spinal cord. It has been proposed (Heffner et al. 1990) that as the large population of cortical efferents descends through the brain stem and successive levels in the spinal cord, the outgrowth of axon collaterals to particular potential target populations of neurons is elicited by the local release of particular chemotropic molecules. An appropriate sequencing of mechanical pathfinder cues to broadly direct the growing axons into the appropriate sites in the brain stem and spinal cord, combined with the local release of specific chemotropic molecules, could efficiently guide individual fibers that originate from particular localized regions of sensorimotor cortex to their respective remote neuron targets.

In comparing the distributions of the corticospinal neuron somas in the newborn and mature macaque, we can see that the areal extent regresses postnatally around its outer rim. Starting medially there is a loss of projection from the whole length of the cingulate gyrus, from the dorsolateral part of area 6aα, from prefrontal area 12, along the lateral borders of areas 6aα and 4, the perinsular cortex, and the intraparietal cortex (including parts of areas 5, 7a, and 7b). In addition, there is a real reduction in the corticospinal soma density throughout the cortical area from which the projection originates in the adolescent and mature macaque (Galea and Darian-Smith 1994, 1995). We (Darian-Smith et al. 1990a,b) have previously observed during the first 8 postnatal months a comparable regression of the thalamic projections to the sensorimotor cortex. During that period, both the areal extent and the projection density of individual thalamic nuclei are substantially reduced. This rough temporal and spatial correspondence in the postnatal regression occurring both in the corticospinal neuron populations and in their subcortical inputs from the thalamocortical neuron populations raises the issue of how the changes in these two populations are related.

2.8 Comment

Recognizing the multiplicity of direct and indirect corticospinal neuron populations in the primate, each transmitting unique information to the spinal cord and each having synaptic access to the spinal circuitry at every segment, has been important in understanding how the cortex regulates movement. This concept has been appreciated mainly in the last decade, its acceptance being boosted by the identification of significant corticospinal projections from the anterior cingulate cortex (Hutchins et al. 1988; Shima et al. 1991). Such connectivity strongly reinforces the view that parallel distributed processing (McClelland et al. 1986) of cortical input to the spinal cord is the rule and that serial processing of sensorimotor information funnelled into a "primary" motor cortex, if it occurs, is only one element of a complex parallel system. Judging from the relative densities of the different inputs to "indirect" corticospinal pathways in the macaque, all of which gain access to the spinal cord through bulbospinal (rubrospinal, vestibulospinal, reticulospinal) projections, the cortical input to each of these systems is at least matched by inputs from the cerebellum, the vestibular apparatus, and bulbar reticular formation. The indirect rubrospinal pathways (see Fig. 2.12), for example, are probably best considered as part of the overall parallel corticospinal transmission system, modulated specifically by the cerebellum.

It will be appreciated that the new experimental data still relate mainly to the macrostructure of these pathways and their cortical and spinal connections. The micro-organization of inputs to these different corticospinal neuron populations is virtually unknown, partly because our definition of the circuitry in any area of cortex is still so primitive. Similarly, details of the spinal terminations of the constituent corticospinal neurons of each of these populations are limited. The best-defined corticospinal terminals are those which synapse directly on spinal motoneurons (for a review, see Fetz 1992; Lemon 1993) and whose somas are in precental area 4. However, our understanding of the operation of these corticomotoneuronal cells in the active monkey or human subject will remain quite limited until the microstructure of spinal segmental and intersegmental circuitry is better defined. Their role in the control of hand/finger movements can only be understood in the context of the operation of the whole spinal circuitry, of which they are but one, perhaps very important, part.

In spite of these current limitations, the immediate future for experimental studies of the functional anatomy of corticospinal neuron populations in the primate brain looks encouraging. When combined appropriately, the new tracer and intracellular labeling procedures are well suited to the analysis of the microstructure of cortical and spinal connections and so far have not been sufficiently used in such studies.

To give these general concepts of the structural organization of corticospinal connections in the macaque more substance, there are now two immediate lines of investigation to be followed using the tracer and labeling techniques that are available. The first of these is to develop a more complete description of the synaptic inputs to individual corticospinal neurons in each of the separate corticospinal populations; this description of inputs should be in terms of the spatial distributions of identified boutons on the soma and dendritic branches of individual corticospinal cells, and the statistics of these synapse distributions for populations of these cells. This data would provide a structural framework for examining the functional integration of these multiple inputs. A second important issue that needs to be elucidated if the operations of the various corticospinal subsystems are to be understood is the organization of the local spinal neuronal circuitry that mediates forelimb and hand movements in the macaque. In spite of its functional importance, studies of this circuitry in the primate using modern anatomical methods have been quite limited.

3 The Primate Sensorimotor Thalamus

3.1 The Thalamus and Sensorimotor Behavior

As Sherrington (1947) described so vividly, all action, including using the hands to carry out highly skilled manipulations, is finally expressed in the ongoing pattern of activity of motoneuron populations which innervate the relevant muscles. This complex patterning of the global motoneuron activity is in turn determined by (a) the input of the different descending projections, both direct and indirect, from the cerebral cortex, as was considered in the previous chapter on corticospinal neuron populations, and (b) the primary afferent input to each spinal segment. Motor action must be played out within the context of a realistic, continuously appraised representation of the physical world within which the subject moves and which includes not only the space external to the subject (the world out there), but also body space, which changes constantly as the different body parts move relative to each other. This implies that visual and proprioceptive and tactile somatosensory inputs to the cerebral cortex are each essential in the formulation and execution of any planned movements. Therefore, any analysis of the neuronal mechanisms mediating, say, the grasping, exploration, identification, and intelligent use of an object within reach must incorporate the neuronal representation of this complex sensory space within which the sensorimotor behavior is acted out. Transferring sensory information from the retina and somatic tissues to the cerebral cortex, although complex in neuronal terms, is conceptually straightforward, and the pathways mediating this transfer have been identified in broad terms, if not in detail, for more than a century. However, the continuous appraisal of this huge inflow of information which specifies the moment-to-moment location of the body parts relative to each other and to the outside world, as well as the functional status of each of the muscles involved, is a far more elaborate process, involving many parts of the central nervous system; not surprisingly, it is still poorly understood. We do, of course, know from clinical and experimental studies of the effects of focal lesions of the parietal and frontal cerebral cortex, of the cerebellum, and of basal ganglia on sensorimotor behavior that these structures have a role in this integrative action.

The aim of this chapter is to provide an up-to-date picture of our present knowledge of the structural organization of some of the neuron populations mediating this sensorimotor integration. It will be appreciated, however, that we still know little of how these different populations contribute to overall sensorimotor behavior. As with the corticospinal neuron populations considered in the previous chapter, the recurring hurdle that currently limits the analysis of sensorimotor integrative function is that, while single neuron recording in the forebrain of the active macaque is well established and of great value, the processing of information by populations of active neurons is still relatively inaccessible experimentally. As was seen in the previous chapter, synthetic models of the representation of sensorimotor information in populations of neurons in the precentral motor and posterior parietal cortex are now being developed (e.g., Georgopoulos et al. 1983; Georgopoulos 1986; Kalaska and Hyde 1985; Kalaska et al. 1989, 1990; Andersen et al. 1992, 1993a,b; Caminiti et al. 1991) and are considered further in the following chapter on the cerebral cortex. Similar studies on thalamic neuron populations are planned for the future.

The thalamus provides the passage of entry for all sensory information to the cerebral cortex, as well as that for all correlated and processed sensorimotor information from the spinal cord, cerebellum, and basal ganglia, including the substantia nigra. In fact, the thalamus provides the cerebral cortex with the key information needed to fully represent both body space and extrapersonal space – the contextual information for all complex motor behavior. Just how this informational transfer occurs is the central problem of thalamic function. For a number of years, a widely accepted model of the organization of thalamocortical projections to the primate sensorimotor cortex was that each nucleus with its own unique subthalamic input in turn projects to a particular region of cortex, which itself receives no other thalamic input (for reviews, see Jones 1985, 1987a–c; Darian-Smith et al. 1990a; Darian-Smith and Darian-Smith 1993). Within this framework, the thalamic nucleus with input from the cerebellum (VLp; Jones 1985; Table 3.1) projects only to cortical area 4, and that with input from the spinal cord (caudal lateral ventral posterior nucleus, VPLc; Olszewski 1952) projects only to the somatosensory cortex. This model of thalamocortical connectivity implies that the thalamus subserves a largely passive relay function and that none of the essential integration of the information from the spinal cord, cerebellum, and basal ganglia that is needed for coordinated sensorimotor behavior occurs until the information from these different sources reaches the cerebral cortex. In this model, each thalamocortical neuron has an exclusive input from only a few functionally similar neurons, and in turn its terminal branches synapse on a quite restricted cluster of cortical neurons, each of which has response characteristics similar to those of the input neurons. This model at first seemed to fit well with other features of the thalamocortical projections. Thus the preservation of mechanoreceptive field characteristics of single neurons in the dorsal column nuclei, VPLc, and the somatosensory cortex, with minimum cross talk between adjacent neurons in the successive nuclei in the pathway (Mountcastle and Henneman 1952; Poggio and Mountcastle 1963), is compatible with a simple relay pathway.

In their thalamic study, Poggio and Mountcastle (1963) made a second observation, subsequently confirmed many times (e.g., Jones and Friedman 1982; Friedman and Jones 1981), which fitted less comfortably with the idea of a simple relay pathway from cutaneous tactile afferents to somatosensory cortex. Thalamic neurons with similar cutaneous mechanoreceptive fields, for example on a contralateral finger, have a roughly parasagittal lamellar distribution extending to the dorsal, ventral, rostral, and caudal boundaries of VPLc (see Figs. 3.6, 3.9). The somatotopic representation of the body surface is in the form of a

Table 3.1. Nomenclature of thalamic nuclei in macaque monkey

Nucleus or nuclear complex (Olszewski 1952)	Subnuclei	Subthalamic input
Anterior	Dorsal (AD)	
	Ventral (AV)	
	Medial (AM)	
Ventral	Anterior (AV)	
Ventral lateral	Oral (VLo)	Pallidal
	Medial (VLm)	Nigral
	Caudal (VLc)	Cerebellar
	Postrema (VLps)	
	X	
Ventral posterior		
Lateral	Oral (VPLo)[a]	
	Caudal (VPLc)	Spinal
Medial (VPM)		Trigeminal
Inferior (VPI)		
Lateral	Dorsal (LD)	
	Posterior (LP)	
Medial dorsal (MD)		
Pulvinar	Oral (Pul.o)	Retinal,
	Medial (Pul.m)	superior
	Lateral (Pul.l)	collicular,
	Inferior (Pul.i)	spinothalamic,
		vestibular
Medial central (CnMd)		
Parafascicular (Pf)		
Paracentral (Pcn)		
Central lateral (Cl)		
Lateral geniculate (GLd)		Retinal
Medial geniculate (GM)		Cochlear

[a]VPLo + VLc + VLps + X (Olzewski 1952) = VLp (Jones 1985).

mediolateral stack of these lamellae within VPLc. The role of such an extensive thalamic neuron population in a simple relay pathway was puzzling and was not resolved. A decade later, using horseradish peroxidase (HRP) to visualize the somas of thalamocortical neurons projecting to the frontoparietal cortex of the macaque, Kievit and Kuypers (1977) approached the problem of thalamocortical organization rather differently. They found that the distribution of somas of thalamic neurons projecting to a localized region of frontal cortex was extensive and lamellar, but commonly not confined to a single thalamic nucleus. There was convergence of input from two or more thalamic nuclei to each small zone of sensorimotor cortex. Later systematic studies, in which the zone of uptake of the retrograde tracer and the distribution of labeled somas in thalamic space were more precisely defined, also demonstrated both lamellar convergence and divergence within the cortical projections from each thalamic nucleus (Ghosh et al. 1987; Leichnetz 1986; Matelli et al. 1984, 1986, 1989; Miyata and Sasaki 1983, 1984; Wiesendanger and Wiesendanger 1985; Wiesendanger et al. 1987; Goldman-Rakic 1987a,b; Darian-Smith et al. 1990a,b, 1993; Rouiller et al. 1994).

What is the function of this structurally elaborate thalamocortical association? The answer must be found in the operations of populations rather than of single thalamocortical neurons. The receptive field characteristics of individual thalamocortical cells will be similar in both the simple passive relay and the convergent/divergent models of thalamocortical transmission. However, these connectivity patterns will differ in the two models. With circuitry mediating a transfer of information by single neurons in much the same form that it is represented in the responses of single cutaneous and other afferent fibers, the emphasis is on restricting both convergence and divergence within each "transmission line" and of sustaining its operation largely independently of other active neurons in the same population of cells. "Surround inhibition" is essential to achieve this. Contrasting with this, in circuitry in which specific information is represented in the coincident activity of many neurons operating in parallel, cross talk between the coactive neurons becomes important. Recent studies of thalamocortical connections in the macaque imply that parallel distributed processing rather than serial processing characterizes the transmission of information in the thalamus.This integrative role for the sensorimotor thalamic nuclei and their connections is more in accord with that proposed for the lateral geniculate nucleus in the visual pathway (Usrey and Fitzpatrick 1993; Merigan and Maunsell 1993) than is the earlier simple relay model.

The above comments alert the reader to the complexities of the architecture of the sensorimotor thalamus and its connections in the primate. The use of the new axonally transported tracers with both light and electron microscopy and intraneuronal labeling with lucifer yellow (LY) have ensured some progress in studying them. Walker's comment a half century ago (1938) that the thalamus "holds the secret of much that goes on within the cerebral cortex" now looks to be a statement that can be examined experimentally, and not just continue to be the expression of a vain hope.

In this chapter, current ideas concerning the functional anatomy of the primate dorsal thalamus and its connections, particularly as they relate to the use of the hand, are examined. As in the previous chapter, the emphasis is on the macaque brain, since this is the one the authors know best. The historical and comparative approaches to the thalamus have been extensively reviewed by Jones (1985) and are not explored in this paper. Jones' monograph provides an excellent vantage point, which we have used extensively, for examining work on the primate thalamus during the last decade.

Few investigators of the thalamus can effortlessly visualize and relate in space the 50 or so irregularly shaped nuclei from the incomplete stacks of coronal or sagittal serial sections that are commonly published in research papers dealing with the primate thalamus. We have been greatly helped by three-dimensional reconstructions of the thalamic nuclei, of their inputs, and of the territorial projections to localized zones of the cerebral cortex of individual experimental animals. This is now quite practicable using readily available commercial software. These reconstructions are derived from 20 or more serial coronal maps of an individual macaque thalamus and can be manipulated, rotated, rendered translucent, and dissected on the screen to great advantage. It is hoped that the stationary figures in this paper will convey some of the insight obtained from the images that can be manipulated on the computer screen and at the same time illustrate some of the problems that arise when attempting to match the topography of particular thalamic connections with the cytoarchitectonic parcellation of thalamic space.

Recent work on the neuronal circuitry in different nuclei in the macaque thalamus is also examined. In addition, the topography and connections of the pulvinar and its relation to both the visual and somatosensory thalamic nuclei are reviewed, since the different neuron populations of the pulvinar must have a special relation to manual dexterity.

3.2 Architecture of the Dorsal Thalamus

3.2.1 Cytoarchitecture: Three-Dimensional Maps of Thalamic Nuclei

The thalamus comprises three parts, namely the dorsal and ventral thalamus and the epithalamus. We focus on the dorsal thalamus, with its mainly sensorimotor role, but consider also the thalamic reticular nucleus, a part of the ventral thalamus which forms a thin shell over the anterior, lateral, and posterior boundaries of the dorsal thalamus and is functionally related to the latter. The thalamus of the mature macaque is about 10 mm rostrocaudally and 6–7 mm across, oriented obliquely, with the anterior poles joined at the midline and the posterior poles displaced laterally from the midline (see Figs. 1.2, 3.2–3.4). The relatively small cluster of anterior thalamic nuclei is at the anterior pole, and medially the quite large mediodorsal nuclear complex extends back for about two thirds of the rostrocaudal extent of the thalamus. The internal medullary lamina separates the medial complex from the quite large ventral and lateral nuclear complexes, whose constituent subnuclei are listed in Table 3.1 and include an expanded pulvinar nuclear group. Perhaps resulting from the expansion of the pulvinar, the posterior nuclear group is relatively small in the primate. The intralaminar nuclei of the macaque are quite large, including rostrally the paracentral and central lateral nuclei (Cl) and caudally the medial central (CnMd) and parafascicular (Pf) nuclei. The medial and lateral geniculate and associated nuclei are included in the three-dimensional reconstruction (Fig. 1.2).

Identifying the separate clusters of neuron somas or "nuclei" in a coronal section of the macaque's thalamus stained with cresyl violet is largely an exercise in distinguishing between visual textures that have been generated by changing mixes of neuron cell bodies of different sizes, shapes, and staining intensities, along with unstained aggregates of axons. If a series of coronal sections of the whole thalamus is systematically examined, about 50 separate nuclei can be identified in the macaque, some being identified in three dimensions with moderate confidence, whereas with others parts of their borders are less certainly specified. Jones (1985) recounts how, in a short period lasting only 20 years starting with Nissl's report in 1889 on the rabbit brain (reported in full in Nissl 1913; see Jones 1985), the major thalamic nuclear groupings that we now recognize were identified in different mammals, including monkeys (Sachs 1909; Vogt 1909; Friedmann 1911). This was well before the connections or other functional characteristics of nearly all of these nuclei were known. Initially there were misunderstandings with nomenclatures, but the early systematic descriptions of the macaque thalamic nuclei soon took on a pattern that is familiar to the modern neuroanatomist (Clark 1932, 1936; Polyak 1932; Crouch 1934; Crouch and Thompson 1938; Walker 1938; Krieg 1948). In 1952, Olszewski published the cytoarchitectonic atlas of the macaque thalamus (Macaca mulatta) that has become the standard reference. In recent years there has been updating of some of the nuclear boundaries defined by Olszewski (Jones 1985; Darian-Smith et al. 1990a), but these changes have been small and were usually prompted by an improved definition and understanding of the connections of the thalamus rather than by improved cytoarchitectonic characterization.

While the cytoarchitectonic nuclear parcellation has provided a useful and reliable topographic map with which other measured thalamic features can be correlated, it has also often helped in demarcating functionally different neuron populations, such as the lateral and medial geniculate nuclei and the somatosensory VPLc (Olszewski's classification of 1952) within thalamic space. The early investigators were convinced of the following (Olszewski 1952):

> The value of architectonic methods lies in the fact that subdivisions based on these methods have biological value. The delineated units have either different connection, or different ontogenetic development, or different reactions toward disease, or . . . , without, however, specifying . . . (their) . . . nature . . . and . . . importance.

Proposed nearly a century ago, with little supporting evidence, this view has been remarkably vindicated. Two- and three-dimensional cytoarchitectonic maps of the thalamus have proven to be of great value in correlating the distributions of terminal ramifications of input axons, of the somas and dendrites of intrinsic and relay thalamic neurons, of single neuron responses to specific physiological inputs, and in comparative studies of the primate thalamus, including that of man (Hirai et al. 1988; Hirai and Jones 1989; Jones 1989). However, as was seen in the previous chapter on cortical connections, since the mid-1980s the use of more sensitive anterogradely and retrogradely transported labels for identifying thalamic connections has prompted a sharper and more critical look at the cytoarchitectonic criteria for demarcating thalamic nuclei in the macaque and other primate species and a questioning of some of the earlier correlations of thalamic "nuclear" structure and function. Now that it is possible to visualize the soma/dendrite patterns of many thalamic relay neurons within a single 400 μm thalamic slice (see Fig. 3.12), the need to determine to what extent individual dendrites respect these cytoarchitectonically specified borders is apparent. Some of these issues are considered in the following paragraphs, but often the present picture is quite uncertain.

In Fig. 3.1, the six coronal sections stained with cresyl violet illustrate typical cytoarchitectonic nuclear patterns which have been used to differentiate the 50 or so nuclei in the mature macaque. The various labels are placed centrally within fairly obvious cytoarchitectonically different zones. Some boundary zones defining these nuclei are sharp, such as those juxtaposed to the internal capsule on the left. However, the cytoarchitectonic demarcation of adjacent "nuclei," such as VPLo (oral lateral ventral posterior nucleus) and VPLc (which do in fact differ greatly in their structural and functional organization), is often indeterminate because of an intervening no-man's-land of histological uncertainty and the "interdigitation" of neurons of the apposed nuclei (Jones 1985), which is often invoked to account for a blurred boundary. In reality, these boundary disputes

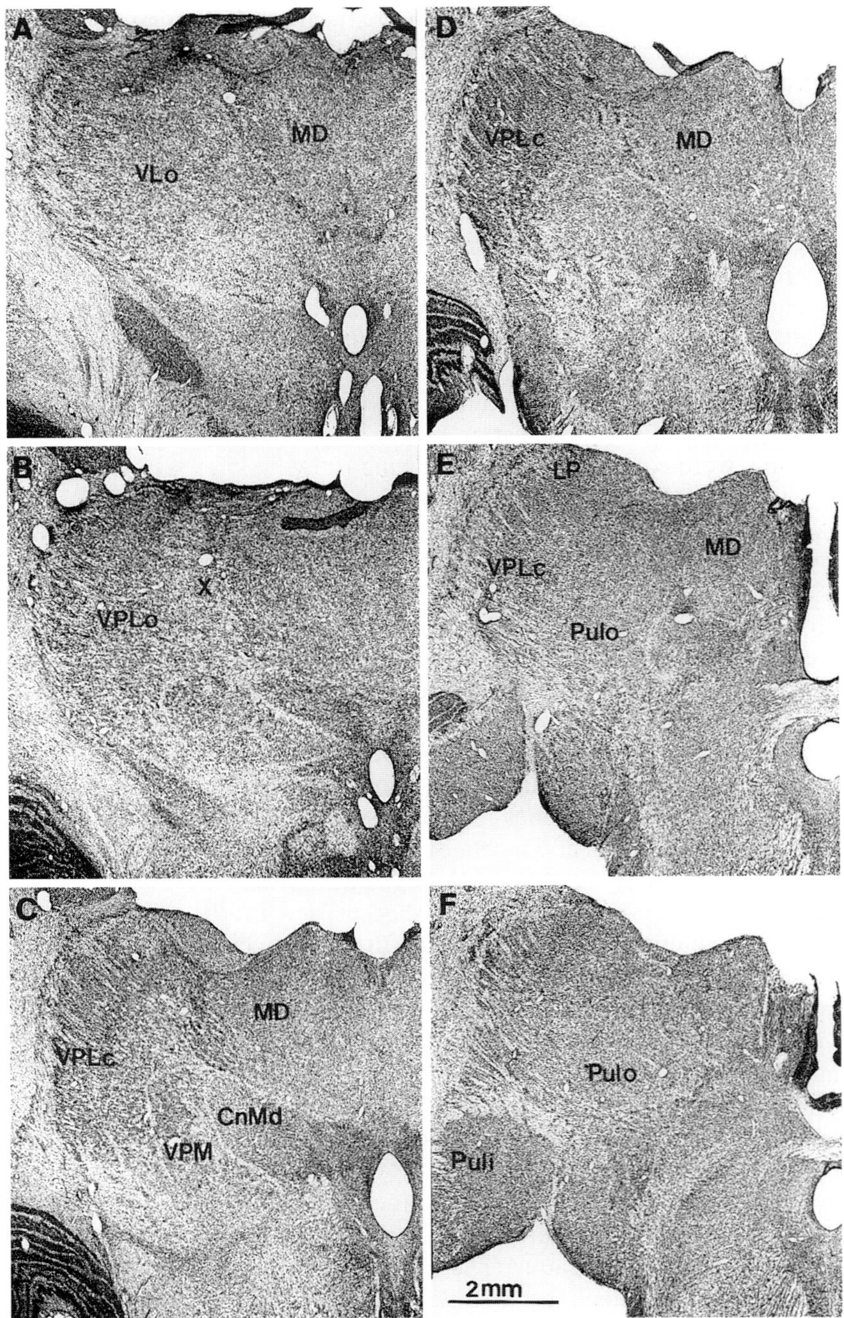

Fig. 3.1A–F. Series of coronal sections of thalamus of *Macaca nemestrina* stained with cresyl violet. Rostrocaudal spread of these sections is approximately 5 mm, extending from the oral ventral lateral nucleus (*VLo*) to the oral pulvinar (*Pulo*), and does not include the most rostral or caudal nuclei. *MD*, medial dorsal nucleus; *VPLc*, caudal lateral ventral posterior nucleus; *VPLo*, oral lateral ventral posterior nucleus; *LP*, posterior lateral nucleus; *CnMd*, medial central nucleus; *VPM*, medial ventral posterior nucleus; *Puli*, inferior pulvinar

have rarely been of moment, since with increasing spatial resolution of the connections and morphology of thalamic nuclear neuron populations it has become clear that commonly there is overlap in their distribution within thalamic space or even a gradual transition of their structure and function. It is ironic that the nuclei labeled by Olszewski as VLc (caudal ventral lateral nucleus), VPLo, and X, which can be differentiated in the cresyl violet section, have been shown to have a common cerebellar input and on this basis have been reclassified by Jones (1985) as part of a single nucleus, VLp (posterior ventral lateral nucleus, which includes VLc, VPLo, X , and VLps). In other words, functional and not cytoarchitectonic criteria were (rightly) used to define the nucleus. Having said this, it should be added that Olszewski's cytoarchitectural maps of the macaque thalamus continue to be a valuable guide defining the main nuclear subdivisions of this complex space; uncertainties occur only when attempts are made to sharply define boundaries with a meaningless resolution.

Figures 3.2–3.4 illustrate the rhesus macaque's dorsal and reticular thalamus constructed from the 21 coronal sections illustrated in Olszewski's (1952) atlas, using AutoCad and 3D Studio software (AutoDesk, Sausalito, California) to examine the thalamus from different viewpoints. We used Olszewski's sections for this three-dimensional construction, since the original coronal Nissl sections (the "raw" data) are readily accessible to the reader. Similar maps were constructed in our laboratory from brains of *M. nemestrina* and *M. fascicularis*; although the brains of these species differ substantially in overall size, the form of each thalamus and its nuclear subdivisions are remarkably similar. It should be noted that some discontinuities occur in Olszewski's series of coronal maps, which have been edited on the basis of our own maps (Darian-Smith et al. 1990a). Olszewski's nomenclature is used in these figures and throughout the review, except for the ventral lateral complex of nuclei, for which we use Jones' logical and simpler parcellation (see Table 3.1). Let us first examine Fig. 1.2, which shows the major cytoarchitectonic nuclei of the macaque thalamus. The enveloping net represents the thalamic reticular nucleus, which is continuous with the perigeniculate nucleus extending over the lateral geniculate nucleus (Lgn). The top figure is of the right thalamus viewed from an anterolateral position; in the middle and lower figures, the dorsal thalamic complex has been rotated clockwise 45° and 90°, respectively. In Figs. 3.2–3.4, the thalamic nuclei are visualized in topographically and functionally related groupings. Figure 3.2 illustrates the topographic relations of the somatosensory thalamic nucleus (VPLc) and the Lgn, which are embedded in the pulvinar nuclear complex, consisting of the oral pulvinar, the medial pulvinar, the lateral pulvinar, and the inferior pulvinar nuclei. The major subthalamic inputs to these nuclei and their cortical projections are also shown. In Fig. 3.3, the more laterally located thalamic "motor" nuclei are shown (VLp, Jones 1985; oral ventral lateral nucleus, VLo; medial ventral lateral nucleus, VLm; and ventral anterior nucleus, VA). Again, the subthalamic inputs from spinal cord, cerebellum, globus pallidus, and substantia nigra to the different nuclei are shown, along with their cortical projections. Figure 3.4 illustrates the main intralaminar nuclei. While these three figures may help the viewer appreciate the topographical relations of the different cytoarchitectonic nuclei of the macaque thalamus, it should be emphasized that the nuclear boundaries are invariably more sharply represented in the figure than is the case. One need only

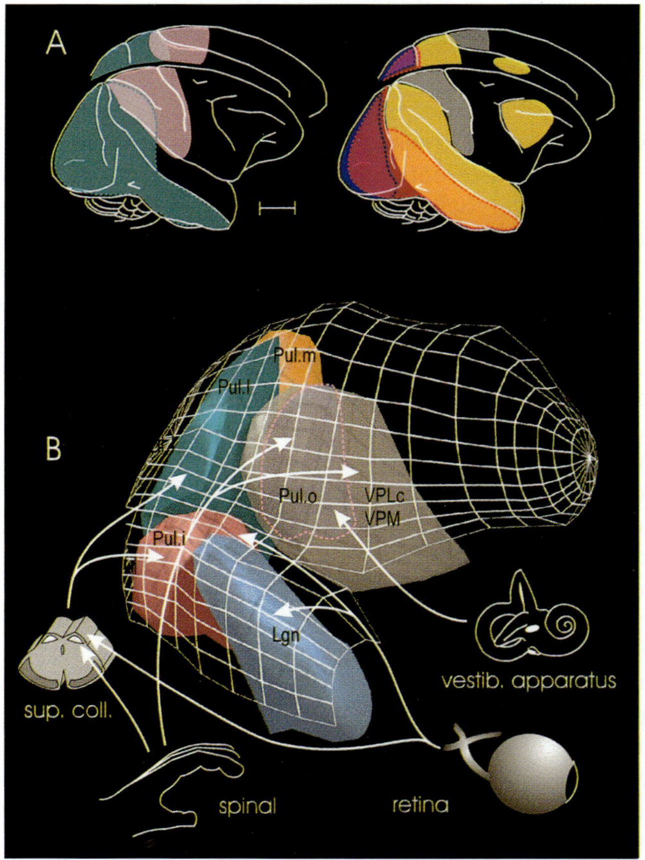

Fig. 3.2A,B. Thalamus of macaque viewed from an anterolateral and slightly elevated position. All nuclei anterior to the caudal lateral ventral posterior nucleus (*VPLc*) have been dissected away, and remaining nuclei are opaque. The VPLc and lateral geniculate nucleus (*Lgn*) are embedded in the more caudal pulvinar complex (oral pulvinar, *Pul.o*; medial pulvinar, *Pul.m*; lateral pulvinar, *Pul.l*; and inferior pulvinar, *Pul.i*). The enveloping net represents the thin thalamic reticular nucleus and its extension over the dorsolateral surface of the lateral geniculate nucleus. The main inputs to these nuclei are illustrated as cartoons, although the cochlear input and the medial geniculate nucleus have been omitted for convenience. The neocortical areas receiving input from each thalamic nucleus are illustrated in **A** and highlight the overlapping of these different projections. The cortical projections are color coded. This three-dimensional image of the thalamus is an accurate reconstruction from coronal sections stained with cresyl violet, similar to those in Fig. 3.1; it is based on several such sets of sections, including Olszewski's published series (1952). *VPM*, medial ventral posterior nucleus (C. Darian-Smith, unpublished)

reexamine the original histological sections in Fig. 3.1 and in Olszewski's atlas (1952) to appreciate this graphical artifact.

3.3 Functional Groupings of Nuclei Within the Sensorimotor Thalamus

Our present understanding of the operations of the different neuron populations and nuclei of the primate sensorimotor thalamus is quite patchy, reflecting in part

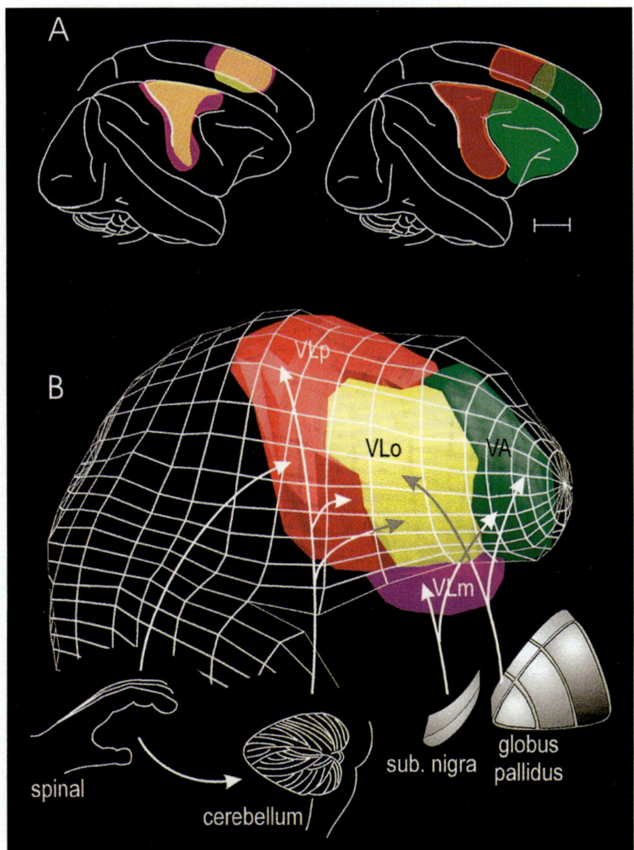

Fig. 3.3A,B. A view of the macaque's thalamus, as in Fig. 3.2, but now the nuclei anterior to the caudal lateral ventral posterior nucleus (VPLc), mostly with a "motor" role, are retained. Again these nuclei are solid, and both their subthalamic inputs and cortical projections are indicated graphically. *VLp*, ventral lateral posterior nucleus; *VLo,* oral ventral lateral nucleus; *VA,* anterior ventral nucleus; *VLm,* medial ventral lateral nucleus (C. Darian-Smith and A. Tan, unpublished)

the fact that one of the important experimental tools, the electrophysiological recording of single neuron responses, as it is currently used is best suited to the analysis of neuron populations with an identifiable direct input from a particular sense organ or receptor neuron population (Fetz 1992). As a consequence, the most intensively studied thalamic nuclei are the visual Lgn and the somatosensory VPLc and VPM ("ventrobasal complex"). Even with these nuclei, however, there is much uncertainty about their functional anatomy and the processing of information by their constituent neuron populations. In order to identify functionally related groups of sensorimotor thalamic nuclei and neuron populations in the macaque, their connections and morphology are examined below.

All nuclei of the primate dorsal thalamus have common structural features on which their particular signature is superimposed. These common features include the following:

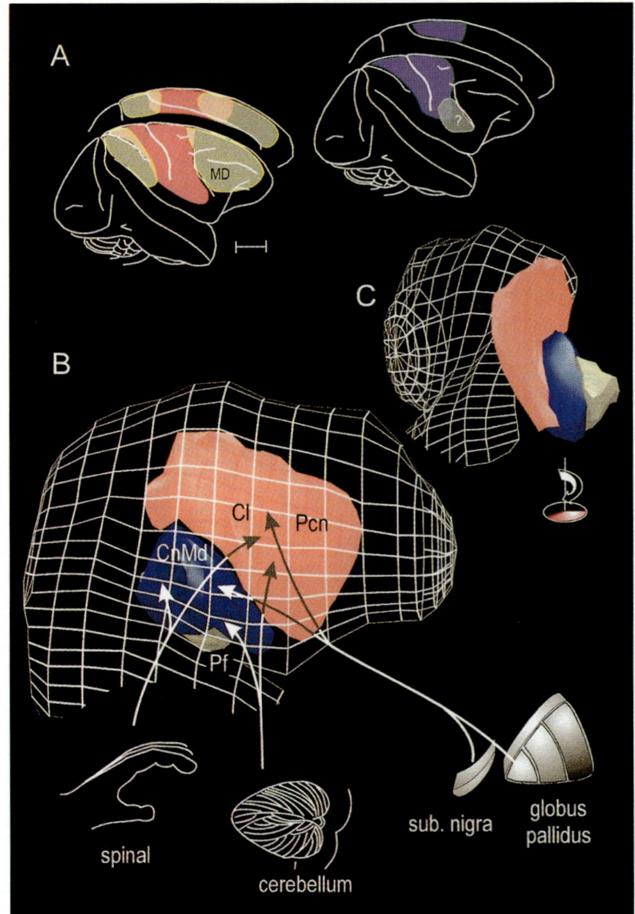

Fig. 3.4A–C. Another view of the macaque's thalamus, but now only the main intralaminar and mediodorsal nuclei are retained. The viewpoint for the central figure is directly lateral, but on the *right* the mediodorsal nucleus is viewed from a directly anterior position. The subthalamic inputs are represented somewhat tentatively. In **A**, thalamic projections to the neocortex are represented and color coded. *Cl*, central lateral nucleus; *Pcn*, paracentral nucleus; *CnMd*, medial central nucleus; *Pf*, parafascicular nucleus (C. Darian-Smith, unpublished)

1. Input from one or more subthalamic neuron populations. For visual and sensorimotor nuclei, these originate from the retina, spinal cord, cerebellum, basal ganglia, or midbrain (for pulvinar). This input may be extensive and complex, as with VPLc, VPLo, or Lgn (for reviews, see Jones 1985, 1987a; Tracey et al. 1980; Usrey and Fitzpatrick 1993; Asanuma et al. 1983a–c; Ilinsky et al. 1993a,b; Macchi 1993), or relatively small compared with other inputs to the nucleus (as in pulvinar and mediodorsal nucleus).
2. Thalamocortical and corticothalamic projections. Every identified nucleus in the macaque dorsal thalamus has a projection to some part of the ipsilateral cerebral cortex (for a review, see Jones 1985), and from this same cortical area a corticothalamic projection originates. These two sets of connections are often loosely described as being "reciprocal" (Jones 1985), a term which says

little about their organization. The thalamic reticular nucleus, in spite of its close topographic association with the dorsal thalamus, is part of the ventral, not dorsal, thalamus and has no cortical connections. Both thalamocortical and corticothalamic axons pass through this outer reticular "shell" and give off collaterals which synapse on reticular neurons. Few, if any of these axons have recurrent terminals in the nuclei of the dorsal thalamus.

3. Local circuit neurons or interneurons which are GABAergic. About 20%–30% of the neuron populations in each nucleus of the macaque's dorsal thalamus have been estimated to be interneurons, whereas in the rat there are apparently no interneurons in the homologous thalamic nuclei (Guillery 1966; Yen et al. 1985b; Spreafico et al. 1993).
4. GABAergic neurons in the thalamic reticular nucleus. These have an orderly axonal projection to adjacent thalamic nuclei (Jones 1975; Houser et al. 1983; Spreafico et al. 1993; Lubke 1993).

Within this context of the shared properties of thalamic neuron populations, attention is now directed at identifying those features which differentiate the various functional groups. We focus on the connections, morphology, and circuitry of the neuron populations that can be separated by their cytoarchitecture in order to establish how these different features are correlated. As described in Chap. 2, recent studies using small libraries of anterograde and retrograde tracers (Wiesendanger and Wiesendanger 1985; Wiesendanger et al. 1987; Leichnetz 1986; Matelli et al. 1984, 1986, 1989; Darian-Smith et al. 1990a,b; Darian-Smith and Darian-Smith 1993; Rouiller et al. 1994; Minciacchi et al. 1986) have reinforced a model proposed by Kievit and Kuypers (1977), in which both orderly convergence and divergence of thalamocortical projections occurs, so that the thalamic projection to a localized zone of cortex may originate from more than one thalamic nucleus. Similarly, the intrathalamic distributions of the terminals of specific subthalamic inputs need to be examined carefully to assess whether one or more thalamic nuclei are targeted. In the macaque, spinothalamic terminal endings certainly diverge to terminate along with other subthalamic inputs in several thalamic nuclei (Berkley 1980, 1983; Apkarian and Hodge 1989a–c). Some investigators (Percheron et al. 1993b) now also question whether the terminal pallidal and nigral projections to the anterior thalamus are segregated.

3.3.1 Subcortical Inputs to the Sensorimotor Thalamus

The main subthalamic inputs to the sensorimotor thalamus of the macaque originate from the spinal cord, the trigeminal nuclei, the deep cerebellar nuclei, the vestibular nuclei, the globus pallidus, and the substantia nigra.

3.3.1.1 Spinal Inputs

Two parallel pathways, the medial lemniscal and spinothalamic neuron populations, transmit tactile and proprioceptive information from each spinal segment to the posterior thalamus. Separate neuron populations also relay

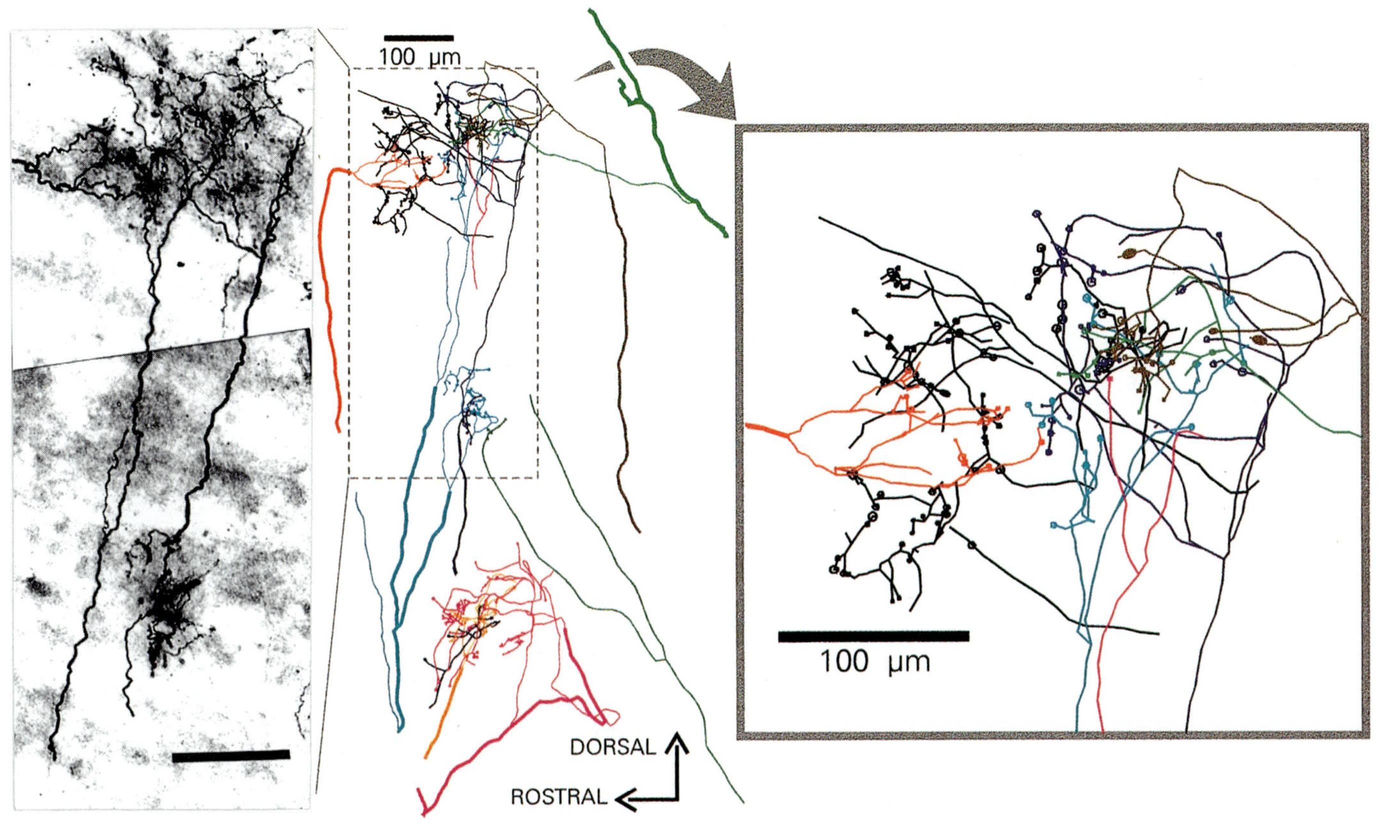
100 μm
DORSAL
ROSTRAL
100 μm

thermoreceptive and nociceptive information in the spinothalamic projection. Each medial lemniscal neuron has a response profile closely approximating that of one of the primary mechanoreceptor afferent types (i.e., Meissner, Merkel, or Pacinian cutaneous afferents or muscle, tendon, or joint afferents). Spinothalamic neurons transmit thermoreceptive and nociceptive as well as somatic mechanoreceptive information from the same spinal segments to a somewhat more extensive region of posterior thalamus (Poggio and Mountcastle 1963; Jones et al. 1982; Willis and Coggeshall 1991).

Medial lemniscal axons projecting from the dorsal column nuclei terminate entirely within the confines of the histologically identified VPLc. This projection is somatotopic, with individual body parts such as the thumb being represented as roughly sagittally oriented lamellae that extend the full dorsoventral and rostrocaudal extent of VPLc; the hand and forelimb are represented medially, and the hindlimb laterally. Individual lemniscal fibers terminate as a localized cluster of boutons within one such lamellar segment (Figs. 3.2, 3.5), extending no more than one third of the dorsoventral and anteroposterior spans of the lamella in VPLc. This terminal branching rarely spreads more than 200 µm mediolaterally (A. Tan et al., in preparation; Jones 1983a). This localized termination of individual lemniscal fibers implies that the "territorial" populations of thalamocortical neurons described later (Darian-Smith et al. 1990a; Darian-Smith and Darian-Smith 1993), many of which occupy a full VPLc lamella, have an input from many of these fibers. In other words, each localized zone (approximately 0.5 mm across) of cortex in the postcentral somatosensory cortex has a convergent input from many medial lemniscal neurons which synapse on relatively discrete clusters of thalamocortical neurons contained within a particular lamellar space in VPLc.

Spinothalamic axons also have a dense terminal projection within VPLc, again as focal clusters or islands. Medial lemniscal and spinothalamic axon terminals originating from the same spinal segments are closely associated in VPLc (Mehler et al. 1960; Bowsher 1961; Boivie 1979; Berkley 1980, 1983; Apkarian and Hodge 1989a,b). Whether any neurons in VPLc receive synaptic input from both medial lemniscal and spinothalamic axons is not known, although some evidence suggest that this may occur (Ralston and Ralston 1994a,b; see below). Spinothalamic projections, however, are not confined to VPLc (and VPI, inferior ventral posterior nucleus). In the lateral thalamus, spinothalamic axons also

◀ **Fig. 3.5.** Lemniscal afferent terminals in the caudal lateral ventral posterior nucleus (VPLc) in the macaque, which have been anterogradely labeled by injecting biotin dextran amine into contralateral cuneate nucleus. The bouton clusters are located in the rostral part of VPLc and are viewed in parasagittal sections. The photomontage on the *left* is taken from a 100-µm section. These axons have been traced using the Eutectic Neuron Tracing System and "merged" with their extensions in two adjacent sections to produce the *middle* diagram (total parasagittal width of section, 300 µm). Boutons of separate axon terminals are color coded, with *black* being those of unidentified origin. *Dashed rectangle* corresponds to photograph on the *left*. On the *right*, upper cluster of boutons has been enlarged to show detail. Note that terminal ramifications with boutons occur in clusters comparable in size with the dendritic tree of a thalamocortical neuron, one lemniscal axon may branch to terminate as two separated clusters, and there is convergent input from multiple axons to each cluster of boutons. *Bars*, 100 µm (A. Tan, unpublished)

terminate caudally in the pulvinar, mainly in the oral pulvinar, and in the suprageniculate nucleus (SG). Rostral to VPLc, spinothalamic axons terminate mainly in VPLo. This extensive rostrocaudal projection to VPLo, VPLc, and the oral pulvinar is continuous. In the medial thalamus, there is a significant spinothalamic projection to the caudal part of the mediodorsal (MD) nucleus and also to the intralaminar Cl and Pf nuclei. The functional implications of this spread of spinothalamic input to the dorsal thalamus are considered later when discussing thalamocortical territorial projections.

3.3.1.2 Cerebellar Inputs

Asanuma et al. (1983a,b) and others (Kalil 1981; Schell and Strick 1984; Rouiller et al. 1994) have used axon tracer techniques to examine the thalamic projections from the deep cerebellar nuclei in the macaque. These projections terminate mainly in the ventrolateral nuclei designated by Olszewski (1952) as VPLo, VLc, X, and VLps (Jones' VLp), but also in the intralaminar nucleus Cl (Figures 3.3 and 3.4) and VLc (Rouiller et al. 1994). Projections from the somatotopically organized dentate and interpositus nuclei to the contralateral thalamus are also somatotopically organized. Each body region is represented in a lamellar space extending dorsoventrally across the nuclear complex. This lamellar distribution is spatially in continuity with the lamellar projection of medial lemniscal input from the same body part to the immediately caudal VPLc. Rostral body parts are represented medially mainly in nucleus X, and caudal parts such as the foot are represented laterally (mainly VPLo). The terminal projections from the dentate and interpositus nuclei are macroscopically aligned, but appear to terminate on separate clusters of thalamic neurons (Asanuma et al. 1983a,b). The less dense thalamic projection from the fastigial nucleus is bilateral, but mainly contralateral; few fastigial fibers terminate in the medially positioned nucleus X (area of hand representation for other cerebellar projections). It will be recalled that each deep cerebellar nucleus has very different inputs from the brain stem and cerebellar Purkinje cells. The input to the dentate nucleus is mainly from the cerebellar hemisphere and from frontoparietal cortex through pontine nuclei and has been proposed as having some role in the planning and initiation of movements (Thach 1978; Brooks and Thach 1981; Thach et al. 1992, 1993). The input to the interpositus nucleus is mainly from spinocerebellar neuron populations and from the motor cortex through pontine relays. A suggested role for this circuitry has been the detection and correction of any mismatch between intended and actual movements (Thach et al. 1992). The fastigial nucleus, with inputs from the brain stem vestibular and reticular nuclei and from the spinal cord, has been thought to be concerned with postural control. Thus the sensorimotor information transmitted from the cerebellum to VPLo, VLc, VLps, and X is both complex and diverse, and probably differs considerably in different parts of this "cell-sparse" (Asanuma et al. 1983a,b) nuclear zone. In spite of this intrinsic complexity of the cerebellar terminations in the thalamus, the ventrolateral thalamic space receiving this input does seem to be sharply segregated from that with input from the medial lemniscus. Spinothalamic terminal projections, however, do not display this sharp separation of the targeted thalamic zone and do overlap with the distribution of cerebellar projections. No such segregation is apparent with the different subcortical projections to the intralaminar nuclei.

3.3.1.3 Pallidal and Nigral Inputs

Input to the dorsal thalamus from the basal ganglia originates from the adjacent globus pallidus and the substantia nigra (Figs. 3.3, 3.4), and until recently (Carpenter 1976; Ilinsky 1990; Ilinsky et al. 1993a,b; Ilinsky and Kultas-Ilinsky 1990; Percheron et al. 1993a,b; Anderson and Turner 1991; Andersen et al. 1993a,b; Jinnai et al. 1993; Rouiller et al. 1994, Kultas-Ilinsky and Ilinsky 1991; Wiesendanger and Wiesendanger 1985; Wiesendanger et al. 1987) its structural organization in the macaque had been neglected. Most of these earlier studies emphasized the discreteness of the terminations of the pallidothalamic and nigrothalamic projections in the primate; further, the Ilinskys considered that they were restricted to specific cytoarchitectonically defined nuclei, with the pallidal projection terminating in the ventral anterior nuclear components VApc (parvicellular) and VAdc (densicellular), and the nigral projection targeting VAmc (magnocellular). This last cytoarchitectonic classification is that of Ilinsky and Kultas-Ilinsky (1987); VApc and VAdc together approximate to Olszewski's VA and VLo, and VAmc to Olszewski's VAmc and VLm. This apparent segregation of subthalamic inputs in the primate contrasts with the substantial overlap of cerebellar, pallidal, and nigral inputs that occurs in nonprimate mammals (Ilinsky et al. 1985). However, with the use of the new anterograde tracer biocytin, Rouiller et al. (1994) found more extensive distributions of the cerebellar and pallidal projections in the macaque than had been described earlier, and these overlapped somewhat. In Olszewski's terms, the pallidal projection was dense in VLo and VA and present but sparse in the adjacent nuclei including VLc, VPLo, VLm, and the more medial MD. Unravelling the confusing topography of these pallidothalamic and nigrothalamic projections in the primate needs further systematic analysis using three or more of the new anterograde labels (e.g. biotinylated dextran amine, BDA; lucifer yellow dextran, LYD; and fluorescein dextan amine, FDA) in each monkey. This would allow the direct comparison of the anatomy of these different projections in thalamic space, without recourse to a cross-referencing to the cytoarchitectonic parcellation, which itself is ambiguous and subject to different interpretations.

As with the spinal and cerebellar projections to the dorsal thalamus, pallidothalamic neuron populations are composite. Cortical input to the striatum occurs from every part of the cerebral cortex and has been pictured as a series of relatively segregated, parallel projections from the sensorimotor, prefrontal, and limbic cortices. Information from each of these cortical regions is processed in the basal ganglia and transmitted in parallel by pallidal neuron populations to the thalamic regions described above. This information in turn is fed back to both the cerebral cortex (Fenelon et al. 1990; Tokuno et al. 1992, 1993) and the striatum (for a review, see Gerfen 1992). Once again, the pallidal inputs to the dorsal thalamus are topographically complex, and their terminations are not strictly bounded by cytoarchitecturally defined nuclear regions. Furthermore, the pallidothalamic projection includes multiple, functionally distinctive neuron populations, each transmitting related but different information concerning sensorimotor behavior. Some idea of the diversity of information relayed to the thalamus by these different neuron populations is evident from the symptomatology of the two common degenerative diseases of the basal ganglia; in Parkinson's disease the initiation of voluntary action is severely impaired, whereas in Huntington's disease movement occurs but is uncontrolled.

3.4 Thalamocortical Neuron Populations of the Sensorimotor Thalamus

The structural organization of thalamocortical projections to the sensorimotor cortex of the macaque is complex. To make the first as yet incomplete step of defining their structure, the following features need to be described systematically:

- The morphology of individual thalamocortical and intrinsic neurons in each nucleus, the identification and soma/dendrite distribution of input synapses on these cells and their ultrastructure, and the connectivity of these cells within the thalamus and in the cortical target areas
- The subpopulations of these cells in each thalamic nucleus
- The topography of the various thalamocortical projections of the neuronal subpopulations within each nucleus, including patterns of convergence and divergence
- The circuitry linking the thalamic reticular nucleus to other thalamic nuclei
- The topographic and synaptic relations of the corticothalamic neuron populations linked with specific subpopulations of thalamocortical neurons in the different thalamic nuclei

We start this section by considering the topography of the connections between the sensorimotor thalamus and cerebral cortex, and we then examine quite selectively the morphology of individual thalamic neurons.

3.4.1 Thalamic Territories

Figure 3.6 illustrates in a coronal map the soma distributions of thalamic neurons, visualized by retrograde labeling, that project to five different and fairly small cortical zones, each less than 1 mm in horizontal extent, in the precentral areas 4, the supplementary motor area (SMA), the postarcuate cortex, and area 6aα; the coronal map illustrates the complexity of the different thalamocortical projections apparent even in a single two-dimensional section (Darian-Smith et al. 1990a). The full extent of any one thalamic projection to a localized zone of sensorimotor cortex can be visualized in serial coronal maps with a rostrocaudal spread of 4–5 mm, but a more informative display of this spatially elaborate thalamic "territory" of labeled neuron somas is provided by a three-dimensional reconstruction, as shown in Figs. 3.7 and 3.8. Figure 3.7 illustrates the soma distributions of thalamocortical projections to medially located zones in cortical

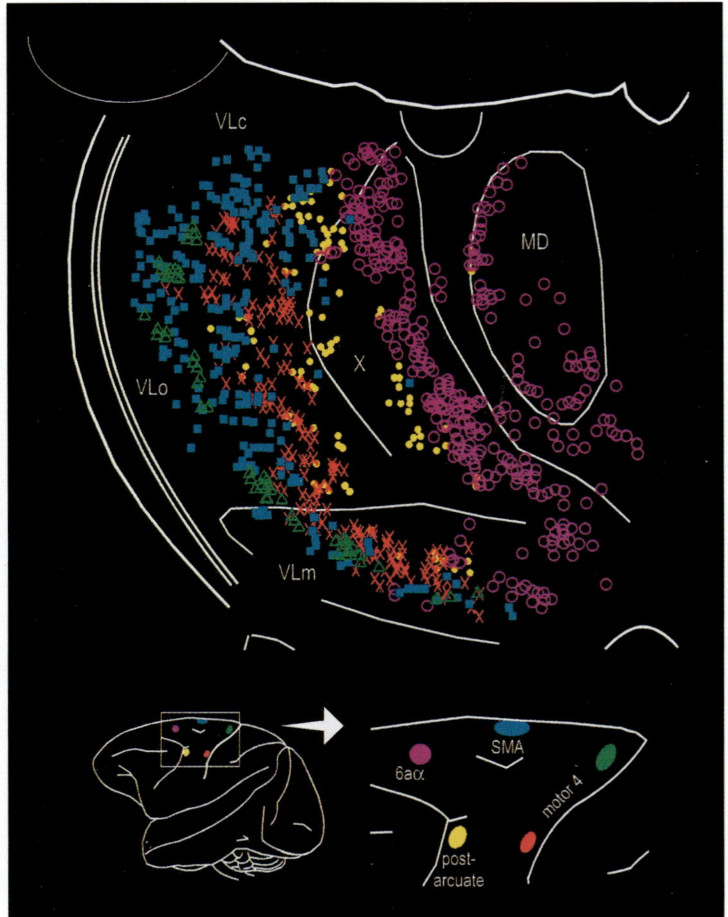

Fig. 3.6. Single coronal map of somas of thalamocortical neurons which were retrogradely labeled following injections of five fluorescent dyes into the sites in the frontal cortex that are shown shown in the *inset*. The lamellar distribution of each labeled population is clearly defined. While these distributions may overlap in thalamic space, it is rare for individual thalamocortical neurons to project to more than one injection site, even when these sites are little more than 1 mm apart (i.e., double-labeling of individual somas was uncommon). This particular map is of labeled cells in an infant macaque delivered at a gestational age of 142 ± 3 days (Darian-Smith et al. 1990a). Series of such maps are used to define the total thalamic space (thalamic "territory") within which these somas are distributed, as in Figs. 3.7 and 3.8. *VLc*, caudal ventral lateral nucleus; *VLo*, oral ventral lateral nucleus; *MD*, medial dorsal nucleus; *VLm*, medial ventral lateral nucleus; *SMA*, supplementary motor area

areas 6aα, the SMA, and areas 4 and 3b (foot representation in primary motor and somatosensory cortex). In Fig. 3.8, the thalamic projections are to the region of hand representation in areas 3b, 4, and postarcuate 6aα, from the back forward. To appreciate the complexity of these different thalamocortical projections and their soma distributions in thalamic space, the relevant thalamic nuclei are first shown, followed by the representation of the territories within a transparent thalamus; finally, the territories are illustrated relative to each other in space, but

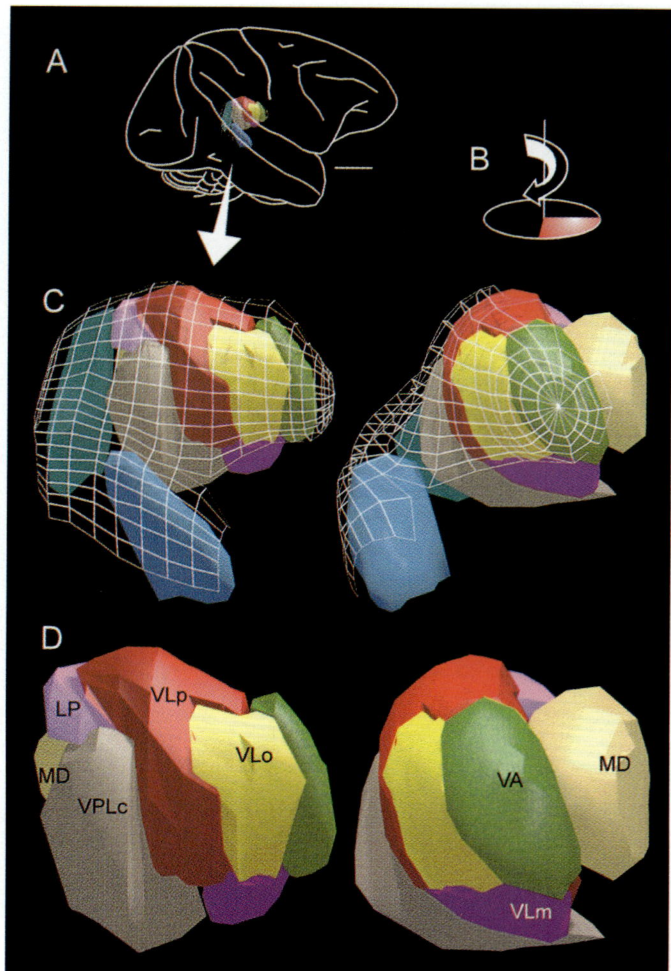

Fig. 3.7A–G. Series of three-dimensional maps of the macaque's right thalamus viewed from both a lateral (*right*) and anterior (right) viewpoint, as indicated in **B**. **A** Thalamus located in the whole hemisphere. **C** Topography of thalamic nuclei (Olszewski 1952) associated with somatic sensorimotor functions, along with lateral geniculate (*metallic blue*) and lateral pulvinar nuclei. *LP*, posterior lateral nucleus; *MD*, medial dorsal nucleus; *VPLc*, caudal lateral ventral posterior nucleus; *VLp*, ventral lateral posterior nucleus; *VLo*, oral ventral lateral nucleus; *VA*, anterior ventral nucleus; *VLm*, medial ventral lateral nucleus. **D** Nuclei through which the thalamic projection territories extend in the series of maps which follow in **E–G**. Thalamic territories are defined as the thalamic space occupied by a continuous distribution of neuron somas projecting to localized zones of cortex less than 1 mm diameter. **E** Four projection zones (i.e., injection sites of retrogradely transported labels) near midline of the territories mapped in **F** and **G**. Thalamic territories and corresponding cortical projection zones are color coded. *Dark blue*, area 5; *green*, area 3b (foot representation); *purple*, area 4 (foot); *yellow*, supplementary motor area, SMA; *red*, area 6aα. **F** Solid thalamic territories viewed through the translucent thalamic nuclei into which they extend. **G** Solid thalamic territories with the nuclei stripped away. Note that for these medially located cortical target zones, lamellar-shaped teritories occupy an outer shell within the ventral lateral subdivision of the thalamus and that the territories projecting to areas 3b (somatosensory "foot" representation) and 4 (motor) are in register rostrocaudally. All thalamic territories cross architectonic borders and often overlap spatially. Thalamic territories in the anterior pulvinar are not shown in these maps. Reconstructions from series of coronal maps, as in Fig. 3.6, were prepared (by C. Darian-Smith) using AutoCad and 3D Studio (Autodesk, Sausalito, California) software

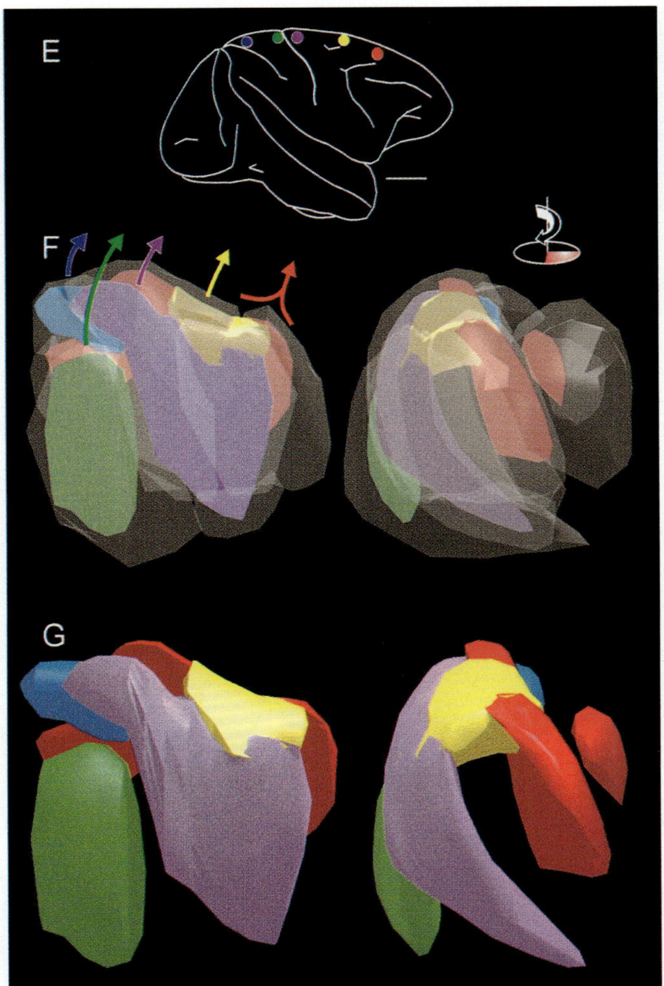

Fig. 3.7E–G

stripped of the surrounding thalamus. For each step in the sequence of reconstructions, the thalamus is viewed from both a lateral position (on the left) and from the front (on the right).

There is thus a substantial convergence of thalamic input to each local zone of motor cortex (approximately 500 μm or less). Although the soma territories of thalamic neurons converging on adjacent cortical zones 1 mm or more apart overlap considerably within thalamic space, there is little sharing of the constituent thalamocortical neurons of these projections. This is demonstrated by the fact that the horizontal spread of the terminal branching of individual thalamocortical neurons is only of the order of 1–2 mm (Garraghty et al. 1989; Darian-Smith et al. 1990a,b). Every small subpopulation of neurons within localized zones of area 4, for example, will therefore have its own unique thalamic input, which transmits a unique mix of information from the deep cerebellar nuclei, from the globus pallidus and substantia nigra, and probably also from the spinal cord. The actual

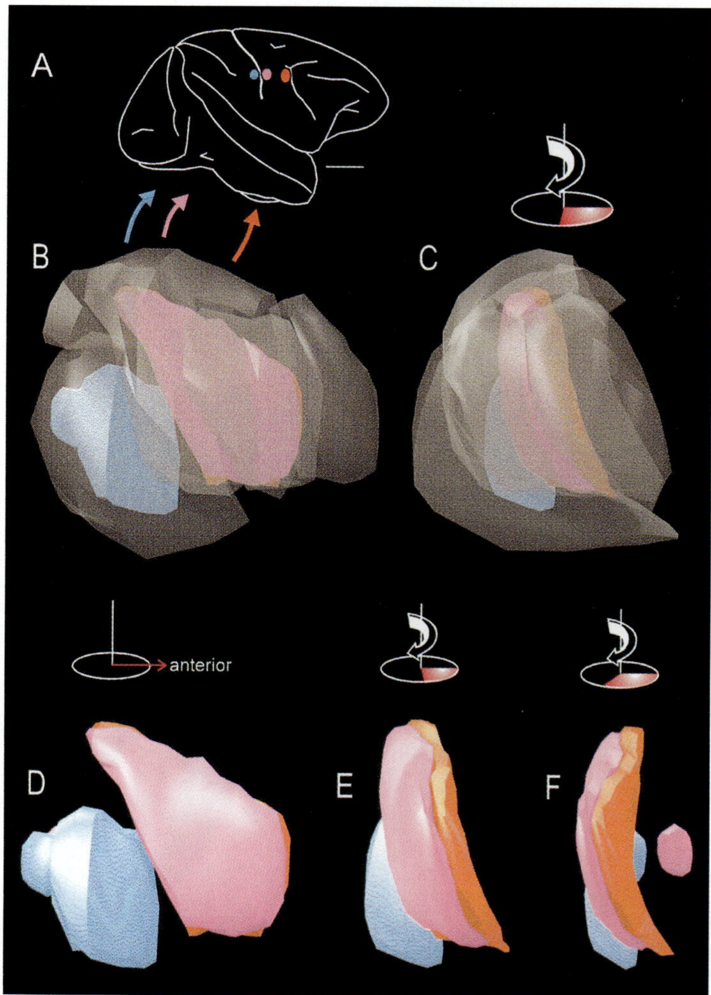

Fig. 3.8A–F. Three-dimensional reconstruction of the macaque's right thalamus and territories, as in Fig. 3.7, but here the retrograde labels were injected into the laterally located region of representation of the hand in the pre- and postcentral cortex (*light blue*, area 3b; *pink*, area 4; *orange*, postarcuate area 6aα). Translucent nuclei in **B** and **C** correspond to those in Fig. 3.7D, but with the medial ventral lateral nucleus (VLm) removed. Territories embedded in these nuclei are medially located. Smaller territories seen in **D** and **F** define small clusters of thalamocortical cells in the oral pulvinar. Although thalamic territories may overlap substantially in thalamic space, they rarely share cells

spatial dimensions of each local subpopulation of cortical neurons receiving a unique thalamic input and extent to which adjacent populations overlap and share neurons is not known. There is, however, little current evidence to suggest that the thalamocortical projection is mosaic or modular in its organization). Figure 3.9 illustrates the essential convergent organization of the thalamocortical territory, in this example projecting to the motor area 4 in the macaque. The neuron populations within each localized region of cortex receive input from more than one thalamic nucleus, which usually means from more than one subthalamic input. Most such territories are lamellar in shape and constitute a continuous segment of thalamic space. However, discontinuities are common, especially when one source of input to the particular region of cortex is from MD, the intralaminar nuclei, or the pulvinar (see Fig. 3.9).

It will be recalled that in the macaque each thalamic nucleus projects to a quite extensive region of cortex (see Figs. 1.3, 3.2, 3.3, and 3.4). Thus, thalamic input to localized regions of cortex is convergent, and cortical projections of individual thalamic nuclei are divergent. This implies that considerable sorting of information received by the thalamus occurs before it is transmitted to the sensorimotor cortex.

Studies of the topography of thalamic territories projecting to different areas of the sensorimotor cortex (Strick 1975, 1976; Schmahmann and Pandya 1990; Darian-Smith 1982; Darian-Smith et al. 1990a, 1991a,b, 1993b) have shown similar patterns to that illustrated in Fig. 3.9 for projections to motor area 4. The differences observed concern the relative contributions from each thalamic nucleus and hence the information from spinal, cerebellar, pallidal, and nigral sources. Figure 3.10 illustrates the different thalamic nuclear inputs to various areas in the frontal and parietal cortex. This anatomical definition of the relative weighting of subthalamic inputs to different local regions of cortex does not imply a fixed pattern for the functional convergence, but rather a limiting structural

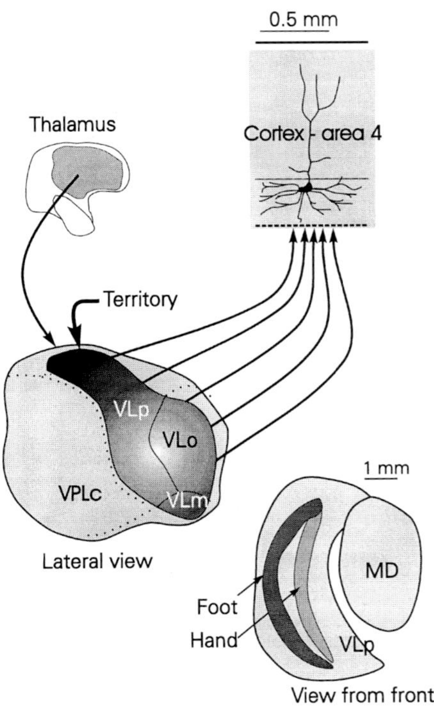

Fig. 3.9. Thalamic *territory* projecting to small zone of area 4 of cerebral cortex of macaque. Axons of thalamic neurons in the *VLp* (cerebellar input) and the oral (*VLo*; pallidal input) and medial ventral lateral nuclei (*VLm*; nigral input) converge to this local zone of cortex. *VPLc*, caudal lateral ventral posterior nucleus; *MD*, medial dorsal nucleus

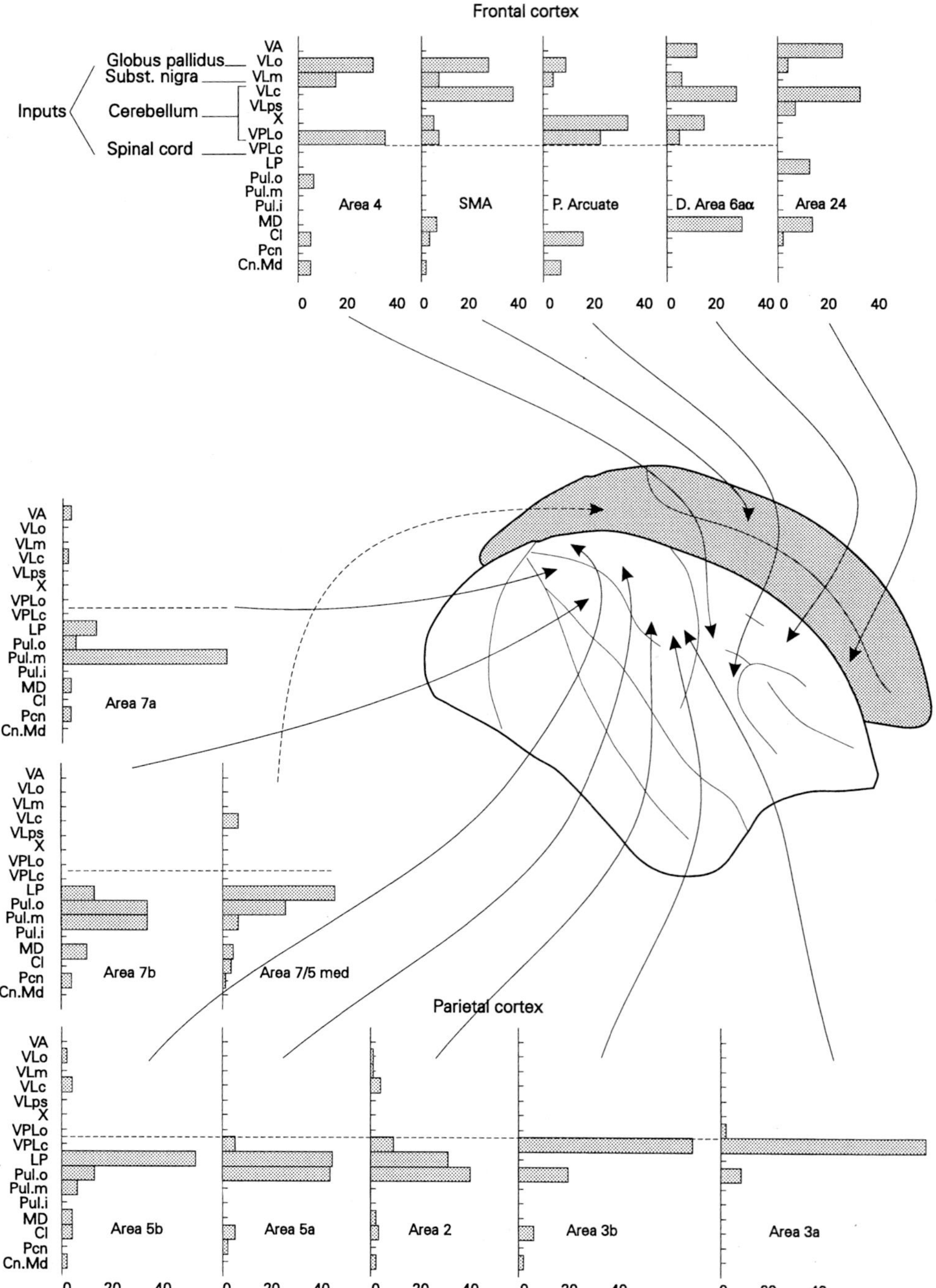

Fig. 3.10. Series of histograms specifying the thalamic nuclear components in the total thalamocortical projection to each of a number of sites in the frontal (*top*), parietal (*lower left*), and mesial cortex in the mature macaque. The simplest mix of nuclear projections is the somatosensory cortex (areas 3a, 3b/1), with a dominant input from the caudal lateral ventral posterior nucleus (*VPLc*) and a small input from the oral pulvinar (*Pul.o*). Moving caudally into the posterior parietal cortex, the VPLc component falls away (areas 2 and 5), and the input from

framework for these inputs. The input from each subthalamic input is likely to vary greatly with the manual task being executed; further, the output of a particular thalamocortical neuron population may be selectively filtered or modulated by corticothalamic or thalamic reticular inputs to it.

One puzzling aspect of the structural organization of these thalamic territories has been that within each thalamic nucleus contributing to a convergent, territorial projection there are many neurons with apparently similar functional characteristics. For example, the constituent cells of neuron populations in cortical area 3b with input from the index finger will have quite localized cutaneous receptive fields on that digit, as will the thalamocortical neurons projecting to this region of cortex. As described earlier, this thalamocortical neuron population is large and distributed within an extensive lamellar zone within VPLc. Why are there so many thalamocortical cells with similar receptive fields converging on one localized population of cortical neurons? The explanation might lie in the fact that information about features of the cutaneous stimulus other than its location on the skin surface (coded by the receptive field), such as tangential movement of a textured surface across the skin, rather than being explicitly represented at the single neuron level is, instead, coded by the population of responding cells. Andersen et al. (1993a,b) have proposed such a distributed representation of several features of the input to a responding population of cortical neurons in the parietal cortex; the representation of information is in the population response, and not evident in the responses of the constituent neurons.

The above description of thalamic territories is based on axon tracer studies and hence is essentially a topographic description. However, to develop a more informative and useful image of the structural organization of this thalamocortical convergence, the resolution must be increased. Some of the necessary experimental steps are as follows:

1. Defining the exact soma/dendrite location of each thalamic neuron with input to a particular single cortical zone
2. Defining the characteristics of the terminal branching of individual thalamocortical neurons and their synaptic relations to the different target cortical neurons in that zone
3. Determining this input pattern for the different neuron types distributed along the cortical perpendicular axis
4. Understanding more fully the roles of each of these cortical neuron types within the elusive circuitry of the cerebral cortex

◀ Pul.o and the posterior lateral nucleus (*LP*) increases; in area 7, the input from the medial pulvinar (*Pul.m*) becomes significant. The parietal cortex also receives a small input from the intralaminar central lateral (*Cl*), and area 7 also has small, but constant inputs from the more rostrally located medial dorsal nucleus (*MD*), and even the caudal ventral lateral (*VLc*) and anterior ventral (*VA*) nuclei. Contrasting sharply with the input to area 3a, the thalamic input to the adjacent motor area 4 is a complex mix from VLp, oral (*VLo*) and medial (*VLm*) ventral lateral nuclei, and intralaminar nuclei. The fractional inputs from these thalamic nuclei to the more rostral parts of the frontal cortex and to the anterior cingulate change in an orderly way. The more rostral frontal and cingulate cortex has a substantial input from MD and small inputs from the intralaminar nuclei. *VLps*, postrema ventral lateral nucleus; *VPLo*, oral lateral ventral posterior nucleus; *Pul.i*, inferior pulvinar; *Pcn*, paracentral nucleus; *Cn.Md*, medial central nucleus. (Constructed from data from Darian-Smith et al. 1990a, 1993b)

Some of these analyses have been started and will be reviewed, but the data are still fragmentary.

3.5 Neuron Circuitry of Sensorimotor Thalamic Nuclei in the Macaque

Current knowledge of thalamic circuitry and processing of the information received is limited. We consider first the topography of thalamocortical projections, followed by the intrinsic circuitry which mediates thalamic processing.

Nearly a century ago, von Kolliker (1896) and Ramon y Cajal (1911) used the Golgi technique to visualize thalamic neurons in the mammalian brain. They described and illustrated three morphologically distinct types in the dorsal thalamus and classified them as follows: (1) neurons projecting to the cerebral cortex, (2) interneurons (both of these found in many parts of the thalamus), and (3) neurons located within the thalamic reticular nucleus, with extensive axonal terminals which spread through other thalamic nuclei. Much of this early description remains valid; thus the thalamocortical relay neurons in the macaque shown in Figs. 3.11 and 3.12 and visualized using intracellular injections of label look similar to the earlier illustrations. In 1966, again using the Golgi technique, Guillery differentiated three types of geniculostriate neuron in the cat's Lgn, which LeVay and Ferster (1977) and others subsequently correlated with geniculate cells that receive input from the electrophysiologically distinctive X, Y, and W retinal cell types. Two decades later, similar form–function correlative studies of thalamocortical neurons in the cat's ventral posterior thalamic nucleus were reported by Yen et al. (1985a,b), but here the morphology of the cell was visualized by the intracellular injection of HRP following its physiological characterization in anesthetized animal (cutaneous mechanoreceptive field, response to joint movements, and identification of axonal projection to the somatosensory cortex by antidromic discharge). Two morphologically distinctive populations of thalamocortical neurons were identified; however, perhaps due to the small cell sample that is inevitable with this difficult experimentation, no functional correlation was demonstrated.

Recently, the in vitro intracellular injection of LY (Buhl and Lubke 1989; Havton and Ohara 1993a,b; Ohara and Horton 1994; I. Darian-Smith et al., unpublished) into retrogradely labeled thalamocortical and corticothalamic neu-

Fig. 3.11. *Top,* thalamocortical neuron located in the caudal lateral ventral posterior nucleus (VPLc) and projecting to zone of hand representation in somatosensory areas 3b/1 in the macaque. This type has a symmetrical, roughly spherical dendritic tree with many thin primary dendrites. The peripheral dendritic branches are "beaded," which may reflect specialization comparable to the dendritic spines of cortical pyramidal neurons. The axon of this cell is not identifiable. *Bar,* 50 μm. *Bottom,* differential labeling of pre- and postsynaptic elements with two-color histochemistry. The VPLc neuron was prelabeled with retrogradely transported fast blue (FB), injected with lucifer yellow (LY), and rendered brown by histochemistry. Corticothalamic boutons (black "tadpoles") were labeled anterogradely with biotin dextran amine (cobalt enhanced). Some of these boutons appear to contact dendrites of the injected thalamic neurons. *Bar,* 20 μm ▶

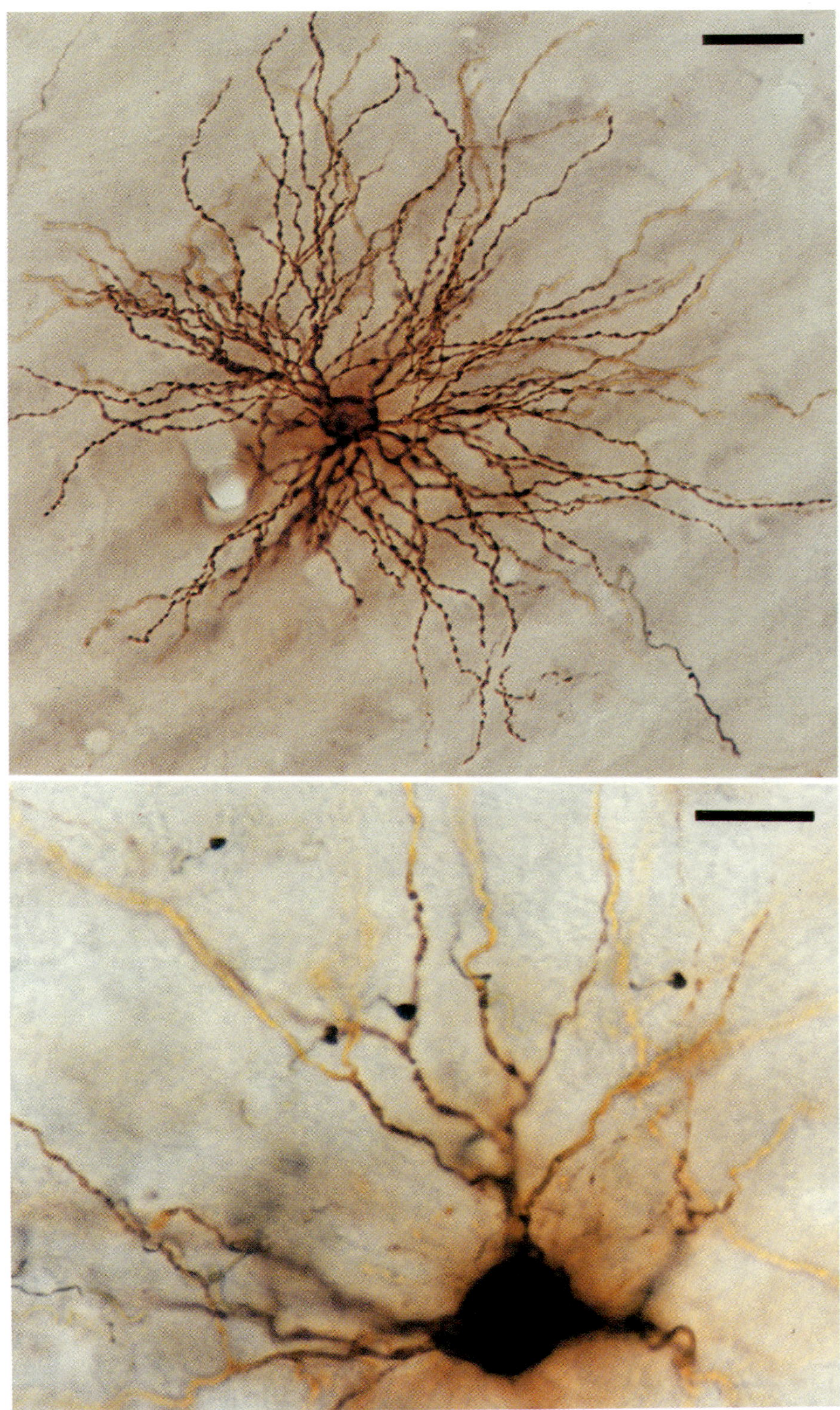

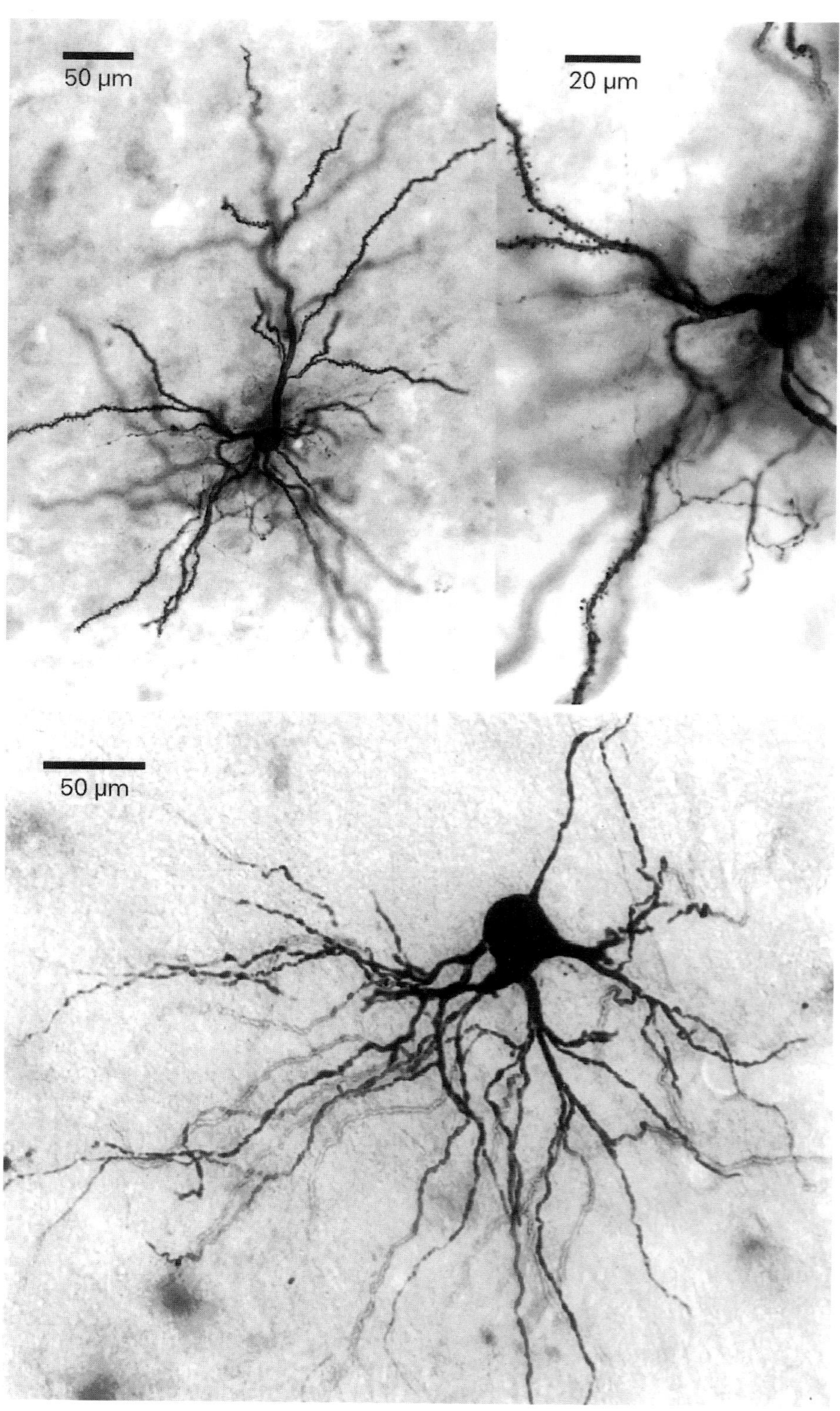
50 µm
20 µm
50 µm

rons has greatly simplified morphological studies of these neurons. Systematic examination of these cells and their connections in the different thalamic nuclei in the macaque is now feasible, but as yet incomplete (Figs. 3.11, 3.12).

3.5.1 Relay Neurons and Interneurons in the Sensorimotor Thalamus

Figures 3.11 and 3.12 illustrate the soma/dendritic morphology of two thalamocortical neurons located in VPLc of the macaque, whose axons project to the hand representation in the somatosensory area 3b/1. The lower cell in each figure has rather stumpy, thick, smooth, tapering primary dendrites which arise from the large soma and then branch into several rather coarse but long secondary dendrites, with a few finer tertiary branches. The thick axon is seen leaving the upper surface of the soma in Fig. 3.12; typically, no collateral branches are evident. These thalamocortical neurons have been called type I cells (Yen et al. 1985a). The second type of thalamocortical neuron found in the macaque's VPLc is shown in the upper part of Fig. 3.11 and corresponds to the type II cell of Yen et al. (1985a). Its soma is somewhat smaller, and many fine primary dendrites radiate out from the soma to form an enveloping, roughly spherical dendritic field. Secondary dendrites arise at bifurcations quite close to the soma, but third-order branching is uncommon. The dendrites are not smooth, but characteristically beaded in appearance, indicating the sites of postsynaptic specialization (Yen et al. 1985a,b). No systematic analysis of the soma/dendritic form of these thalamocortical neurons in the primate has been reported. Yen et al. (1985a,b) concluded that 2 morphologically distinct populations occur in the thalamus of the cat.

In the macaque, the axons of both types of thalamocortical neurons found in VPLc pass dorsolaterally to the internal capsule without giving origin to intrathalamic collaterals; most of these fibers, however, apparently give off collaterals as they traverse the thalamic reticular nucleus (Jones 1985). The terminal dense branching of these same thalamocortical fibers, visualized by the intra-axonal injection of HRP (Garraghty and Sur 1990), is largely limited to lamina 4 and to a lesser extent to laminae 6, in cortical areas 3b and 1, with a horizontal or intralaminar spread varying from 400 to approximately 2500 µm. En passant boutons are densely packed over these terminal branches.

Although Golgi-stained intrathalamic interneurons (Cajal 1911; Guillery 1966) and a few cells labeled by the intracellular injection of HRP (Yen et al. 1985b) have been described, their identification has been equivocal, due to the absence of a labeled axon leaving the thalamus or to the inability to antidromically activate the cell by electrically stimulating the somatosensory cortex. These putative interneurons had a small soma with extensive, irregular dendritic

◀ **Fig. 3.12.** *Top,* small layer VI pyramidal corticothalamic neuron in lateral postcentral cortex of macaque monkey. Cell was first retrogradely labeled with a thalamic injection of fast blue (FB) and then by intracellular injection of lucifer yellow (LY) in a fixed slice of cortex. This is the cell in Fig. 3.13A. On the *right,* the cell's axon with recurrent collaterals is shown enlarged; dendritic spines are also identifiable. *Bottom,* thalamocortical neuron in the caudal lateral ventral posterior nucleus (VPLc), identified as in Fig. 3.11. The morphology of this relay cell differs markedly from those in Fig. 3.11, with only a few primary dendrites, an asymmetric dendritic tree, and an axon originating from part of the soma that is devoid of dendrites (upper part of soma)

branching. Yen et al. (1985b) were unable to identify labeled axons for these cells, although interneurons with axons have been described in the cat's Lgn (Guillery 1966). These cells need to be systematically examined in the monkey before they can be confidently distinguished by their form from relay neurons. However, the occurrence of interneurons in the macaque's thalamic nuclei (but not in the rat) has been established by studies of the ultrastructure of synaptic connections (see below), and the distribution of GABAergic neurons (Spreafico et al. 1993) within them; these cells were estimated to constitute about a quarter of all thalamic neurons.

3.5.2 Neurons of the Thalamic Reticular Nucleus

The thalamic reticular nucleus forms a rather thin cap over the rostral, dorsal, and lateral boundaries of the dorsal thalamic nuclei, intervening between them and the internal capsule; because of this location, it is traversed by all thalamocortical and corticothalamic axons. Its constituent neurons are GABAergic (Jones 1975, 1985; Houser et al. 1983; Spreafico et al. 1993). Cajal (1911) and subsequent investigators (Jones 1975; Yen and Jones 1983; Yen et al. 1985a,b; Spreafico et al. 1993; Lubke 1993) recognized that the soma/dendrite patterns and axon distribution of neurons within the thalamic reticular nucleus differed from those in other nuclei of the dorsal thalamus. Using intracellularly injected LY in nonprimate mammals, Lubke (1993) recently observed a common though variable neuron morphology, with ovoid or elongated neuron somas giving origin to three to six short-stem dendrites. The secondary and tertiary long dendrites run largely within the nuclear sheet; spines and varicosities are seen on these dendritic branches. The axons of these thalamic reticular neurons appear to have a quite orderly distribution in the adjacent nuclei of the dorsal thalamus, although their terminal branching and synaptic organization has not been well described. Axons arising from a particular sector of the reticular nucleus travel with those corticothalamic and thalamocortical axons that traverse that sector and give off collaterals to it. The reticular and corticothalamic axons terminate in the same lamellar zones in which the somas of the thalamocortical cells are located (Jones 1975, 1985; Liu et al. 1995a,b; Mitrofanis and Guillery 1993). In most previous studies of thalamic reticular neurons, nonprimate mammals have been used; since the intrinsic organization of nuclei of the dorsal thalamus differs greatly among species (e.g., a greatly varying interneuron content), these findings need to be verified in primate species.

Clearly, the location and connections of the thalamic reticular nucleus (see Fig. 3.15) imply some form of modulatory function relating to thalamocortical (and thalamostriatal) interactions (Steriade et al. 1988). Both thalamocortical and corticothalamic axons in traversing the reticular nucleus give off collaterals which terminate on GABAergic reticular neurons. These synapses are structurally excitatory (Williamson and Ralston 1993). This implies that the thalamocortical collaterals form part of a recurrent inhibitory circuit feeding back onto thalamic neurons, whereas the corticothalamic axon/GABAergic reticular sequence constitutes a feedforward inhibitory pathway (Williamson and Ralston 1993; Asanuma 1993). Over the last 30 years, some quite elaborate and interesting proposals have been made which attempt to relate this "modulatory" circuitry to various behavioral states, including sleep and selective attention (McCormick 1992; Steriade et al. 1993a,b).

3.5.3 Corticothalamic Neurons

All nuclei of the dorsal thalamus in the macaque have inputs from particular subthalamic sources and from the thalamic reticular nucleus. Each nucleus also has its own unique projection to the cerebral cortex, and in turn a thalamic projection originates from every such area of neocortex. These thalamocortical and corticothalamic nuclear projections have long been regarded as "reciprocal" in the sense that the thalamic target of each corticothalamic projection matches topographically the zone from which the thalamic projection to that cortex originates. Broadly this is so, but with more detailed analysis of the constituent thalamic and cortical neuron populations it becomes obvious that this concept of a reciprocal circuitry is simplistic.

Recent analysis of the morphology of corticothalamic neurons projecting to VPLc in the macaque, using the in vitro intracellular injection of LY and biocytin for visualization (A. Tan et al., in preparation), focuses on some of the complexities of thalamocortical connections. Figure 3.13 illustrates the distribution in pericentral cortex of the labeled somas of corticothalamic neurons projecting to the junctional region of thalamic VPLc and VPLo in the macaque; most of the corticothalamic neuron somas are in lamina VI, but particularly in motor area 4 a significant fraction is in lamina V. The dendritic patterns revealed by intracellular labeling of some of these neurons are also shown in Fig. 3.13. These pyramidal cells had the characteristic apical and basal dendrites, with some apical dendrites extending through all cortical laminae, but with others, although well filled, being limited to the deeper laminae. Many of these corticothalamic neurons had extensive recurrent axonal branching in the deep cortical laminae.

Recently, Tan and coworkers (in preparation) have used mixtures of anterograde and retrograde tracers injected into the pericentral cortex of the macaque to examine both the thalamocortical and corticothalamic neurons associated with a particular localized cortical zone. At a "coarse-grained" level, there are reciprocal projections between zones of cortex (approximately 1 mm across) and crescentic territories of thalamus. However, closer examination reveals a more complex organization (Fig. 3.14). There is almost certainly no one-to-one, exclusively reciprocal, relationship between individual neurons. Many corticothalamic neurons with their somas in layer VI, and a smaller population whose somas are located in layer V, are found in all areas of pericentral cortex. The organization of these two corticothalamic neuron populations, and their thalamic projections appear to vary systematically in the different cortical areas (Giguere and Goldman-Rakic 1988; Velayos et al. 1993; Usrey and Fitzpatrick 1993; Rockland 1994; Yeterian and Pandya 1991, 1994).

3.5.4 Distributions of Terminal Axons Within Thalamic Nuclei

The main axons terminating within each dorsal thalamic nucleus are those originating from subthalamic sources, from the cerebral cortex, and from the thalamic

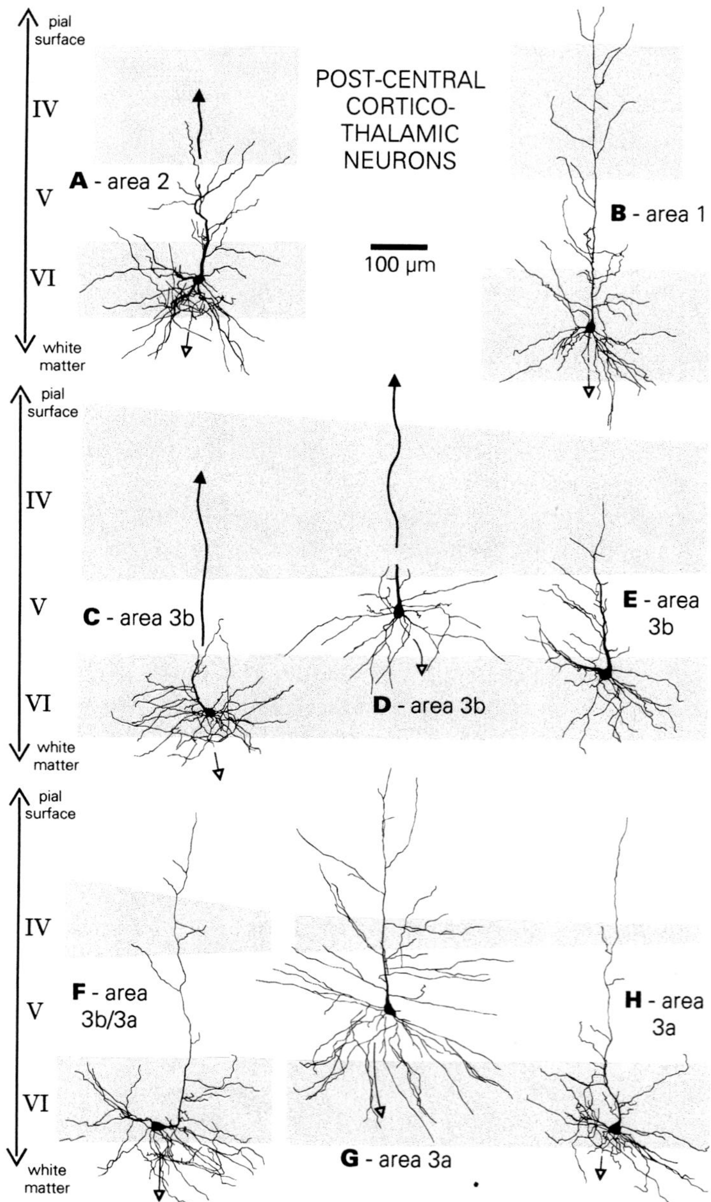

Fig. 3.13A–N. Dendritic morphology of corticothalamic neurons in the hand region of representation in the precentral cortex of the macaque. Corticothalamic somas were first labeled by injecting a retrogradely transported label (fast blue, FB) into the caudal lateral ventral posterior nucleus (VPLc) and the adjacent oral lateral ventral posterior nucleus (VPLo) 2 weeks prior to perfusion of the brain. Slices 400 µm thick were then cut and corticospinal somas visualized using fluorescent microscopy. Individual cells were then impaled with a micropipette, and a lucifer yellow (LY)/biocytin mix was injected iontophoretically. After immunohistochemical conversion of the fluorescent label to a permanent marker, the geometry of the cell was plotted

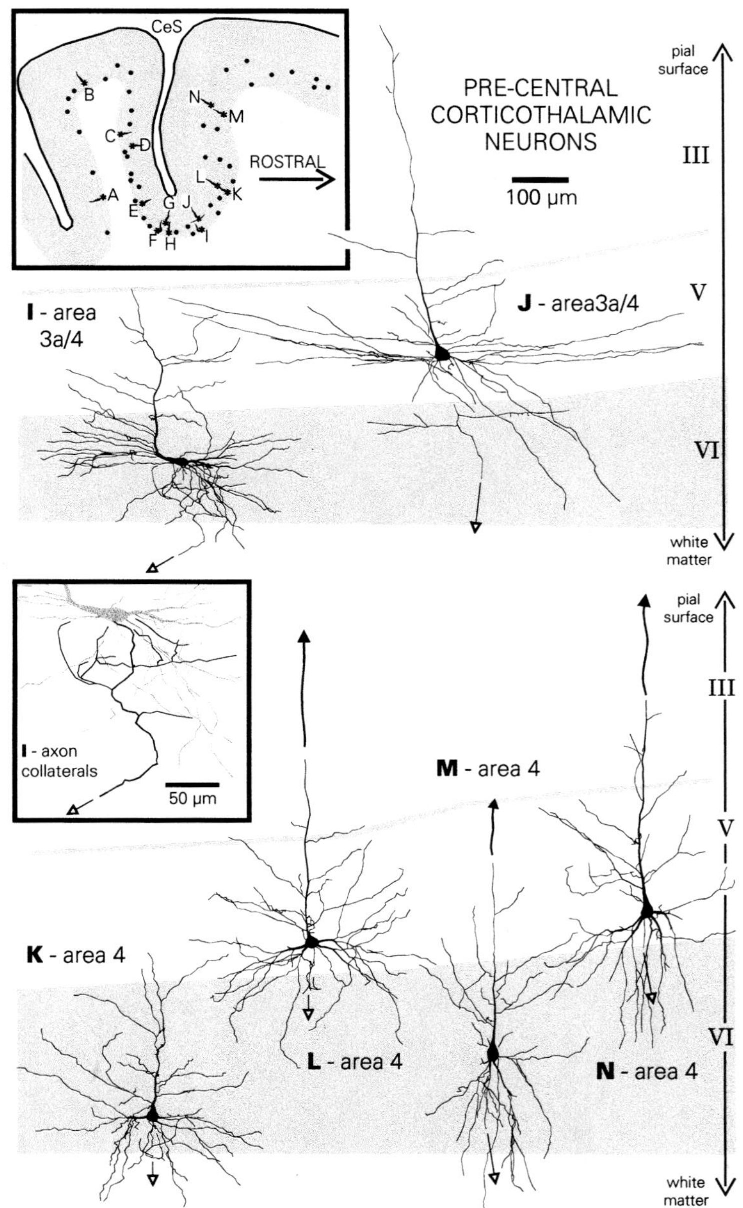

using computer-assisted procedures (Eutectic NTS). In the slice illustrated, 14 cells were labeled. *Inset* at upper right maps the locations of individual injected and other labeled corticothalamic cells in the slice of cortex, which was cut at right angles to the central sulcus (*CeS*). The somas of corticothalamic cells visualized were mostly in cortical layer VI, but some were in layer V. *Upward pointing arrows* indicate that the apical dendrite was truncated by the section; *downward pointing arrows* indicate identified axons. Recurrent collateral axons were commonly filled, and some are plotted in this figure. Dendritic spines were common, but not illustrated at this magnification (A. Tan and I. Darian-Smith, unpublished)

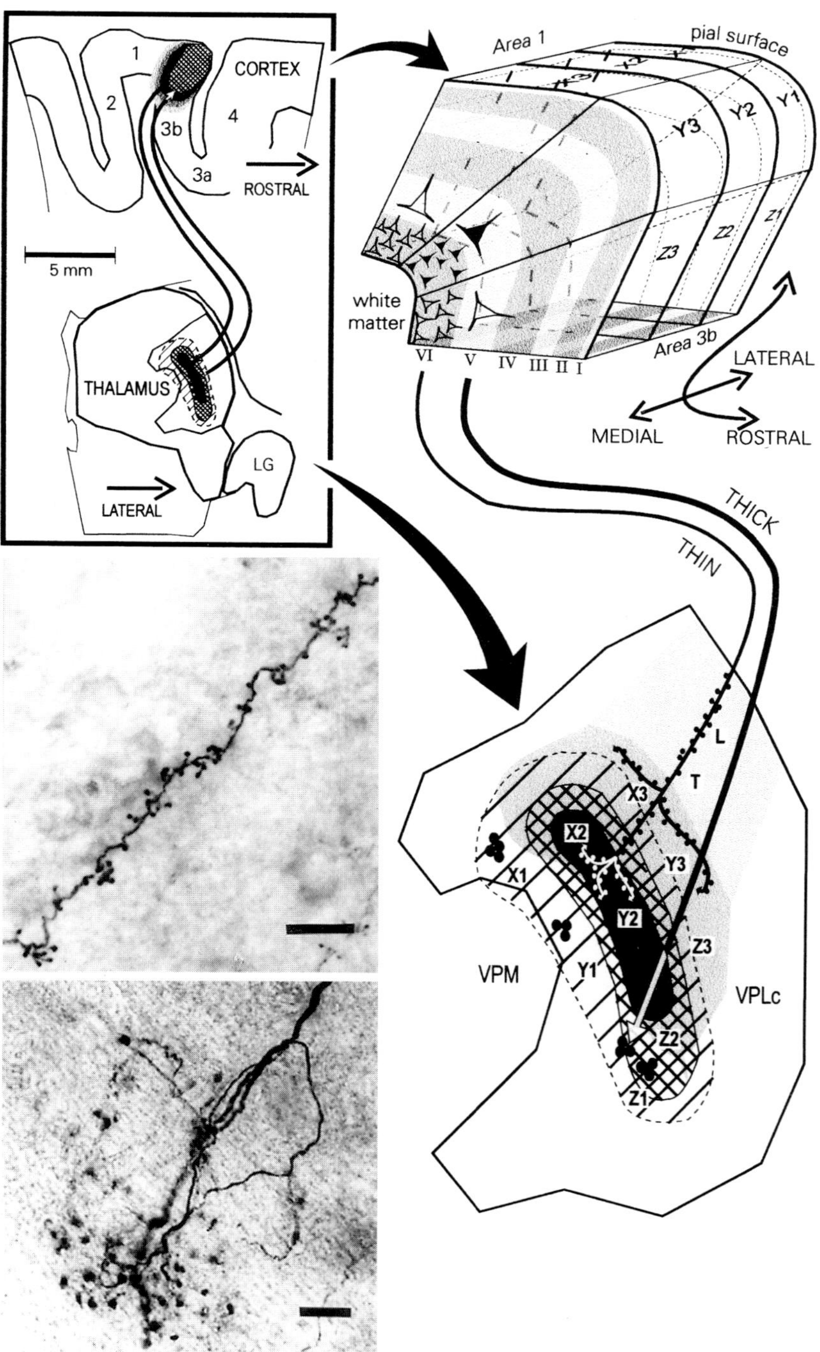

Fig. 3.14. Corticothalamic connections. *Upper left*, neuronal labeling effected with a "cocktail" injection of a retrograde and an anterograde label into somatosensory cortex. Both thalamocortical and corticothalamic cells are labeled, but their relationship is topographically quite complex and is shown in simplified form in the *lower right* diagram. Two types of

reticular nucleus. Whether all or even some thalamic local circuit neurons have axons is uncertain, since a strong case has been put (Ralston 1971; Williamson and Ralston 1993) that these cells synapse with adjacent neurons mainly by presynatic dendritic (not axonal) terminals. The topography of the various axonal arborizations in sensorimotor thalamic nuclei in the macaque has been mainly studied in VPLc. The subthalamic inputs to VPLc are the medial lemniscal and spinothalamic fiber populations. Earlier we described the crescentic thalamic territories and lamellae in VPLc, which define the spatial distributions of somas of thalamocortical neurons projecting to part or whole of the cortical area of representation of a particular body part, such as the pad of the contralateral index finger; they specify the topography of the output of the nucleus. Rather coarse medial lemniscal fibers with mechanoreceptive input from the same finger enter the lamella along the caudal boundary of the nucleus, then pass forward through the nuclear lamella (Figs. 3.5, 3.6), divide into two to three preterminal branches, and then into terminal branches on which many boutons are clustered. These lemniscal terminals are spread through an area of about 0.5×1.0 mm within the nuclear lamina. The latter is somewhat larger (~4 mm rostrocaudal, 3 mm dorsoventral, 0.3–0.4 mm wide), typically extending to the boundaries of VPLc in a roughly parasaggital plane. (Jones 1983c; A. Tan et al., in preparation). Thus each medial lemniscal axon terminates within a relatively localized zone of the respective territory and lamella. The terminal branching of spinothalamic axons is somewhat more expansive (see Apkarian and Hodge 1989c), with concentrations of terminals in VPLc, but also in VPLo and the VPLc/VPLo border (Berkley 1983).

The new anterograde tracers have revealed morphological diversity in corticothalamic axonal projections. Two classes of axons, each with a distinctive type and arrangement of boutons, have been demonstrated in several thalamic nuclei in the macaque (Rockland 1994; Tan et al., in preparation), in the cat (Ojima et al. 1991), and in rodents (Rouiller et al. 1993, 1994). In the macaque VPLc, the more numerous, fine corticothalamic axons ramify densely, with small en passant boutons, in a region coincident with the "reciprocating" thalamocortical territory. However, surprisingly, the main trunk of most of these axons as they stream through the more lateral VPLc toward the reciprocating territory has many short side branches ending in boutons (Fig. 3.14, middle photograph). The other, less common type of coarse corticothalamic axon has no

◀ corticothalamic axons can be identified morphologically. Thick axons terminate in clusters of boutons (*large black dots*) and are seen in the *bottom left* photograph; the evidence suggests that their cell bodies are in lamina V of the cortex. Broadly speaking, these boutons are distributed in a thalamic space enveloping the soma/dendrites of thalamic neurons terminating in the cortical injection site, i.e., near the labeled thalamocortical somas. The commoner type of corticothalamic axon seems to originate from cortical lamina VI, and it does have terminals within the thalamocortical "territory" described above. However, what is striking is the distribution of many stalked boutons along the main course of the axon as it traverses the thalamus from the internal capsule to the targeted thalamic territory. Presumably these boutons synapse on a vast number of thalamic cells before reaching their final targeted population of neurons. The *upper left* photograph illustrates this pattern of boutons. *VPM,* medial ventral posterior nucleus; *VPLc*, caudal lateral ventral posterior nucleus. *Bars*, 25 μm

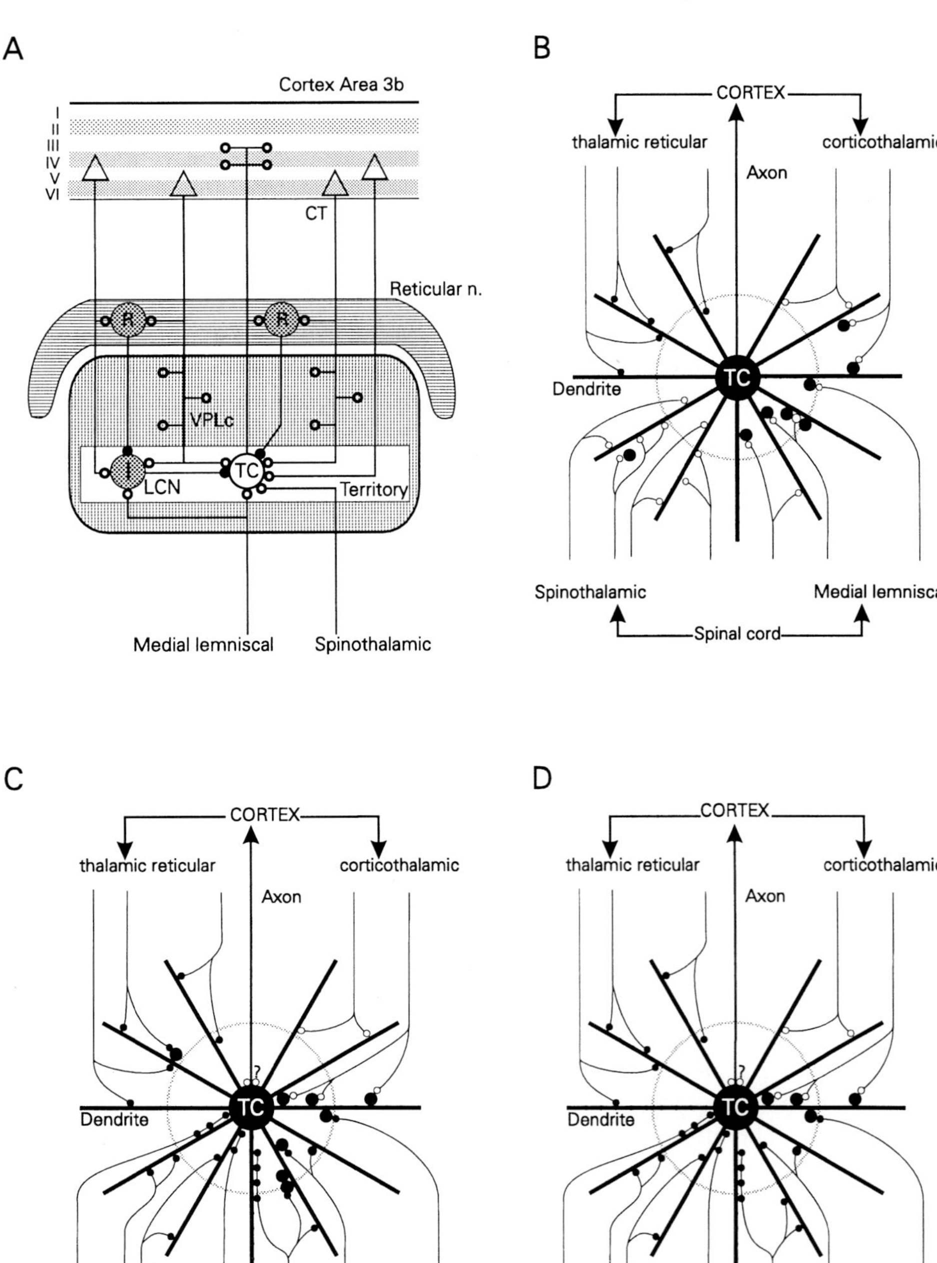

Fig. 3.15A–D. Current views of synaptic connections of different thalamic nuclei in the macaque. **A** Thalamocortical circuitry of somatosensory caudal lateral ventral posterior nucleus (VPLc). Medial lemniscal and spinothalmic projections are the main subthalamic inputs, with terminals from hand (for example) spread through lamellar space (labeled "territory"). Both

side branches and ends in a tight cluster of larger terminal boutons (Fig. 3.14, lower photograph) close to the reciprocating thalamic territory. Thus the fine axons would appear to synapse on thalamic neurons in an extensive part of VPLc adjacent to and including the thalamic territory targeted by their terminal branches, whereas the coarse corticothalamic axons have a much more restricted input to closely adjacent neurons within the reciprocating thalamic territory. The widely distributed synaptic terminals of the fine corticothalamic neurons suggests a modulatory action on a large population of cells within the receiving thalamic nucleus, whereas the targeting of cell clusters within or near the reciprocal thalamic territory by the coarse axons implies a more specific feedback from a quite restricted zone of cortex. Circuitry of this type suggests some form of gain control, but more detailed information of the connectivity within the thalamus is needed to elucidate any such function.

3.5.5 Synaptic Organization Within Thalamic Nuclei

The ultrastructure of synapses in the macaque's sensorimotor thalamus has been more fully examined in VPLc than in any other nucleus. Figure 3.15 illustrates the major connections of VPLc that have been described in the above paragraphs (Fig. 3.15A), together with the current model of locations of the synapses within this nucleus (Fig. 3.15B), which has been largely elucidated by electron microscope studies of prelabeled neurons (Ralston 1971; Ralston and Ralston 1992, 1993, 1994a,b; Ohara et al. 1989). In Fig. 3.15B, a thalamocortical neuron is represented with a central soma (thalamocortical neuron, TC) from which the dendrites radiate to define a roughly circular symmetrical dendritic field. Den-

◀ these subthalamic populations have synaptic terminals on the two cell types, the thalamocortical "relay" cells (*TC*) and GABAergic local circuit neurons (*LCN*) in VPLc; approximately 90% of spinothalamic terminals are on the dendrites of TC cells, whereas most lemniscal afferents terminate on thalamic LCN dendrites, which in turn synapse on TC cells ("presynaptic" dendrites). TC cells project mainly to localized zones in the middle layers of cortical areas 3a, 3b, 1, and 2. Collaterals of TC axons terminate in the thalamic reticular nucleus, but are not recurrent to VPLc. Thalamic reticular neurons, which are GABAergic, project to the adjacent VPLc with inhibitory terminals on both TC and LCN. Corticothalamic neurons (*CT*) seem to consist of two populations with their somas in cortical layer V (few) and VI (many), respectively. Axons of both TC populations have dense terminal branching in the region of the thalamic territory of TC reciprocating the cortical projection. The commoner CT neurons are characterized by many stalked, probably excitatory, boutons along their whole extent. **B** Synaptic input onto a TC relay cell in VPLc. *Open circle*, excitatory terminals; *small filled black circles,* inhibitory synapses; *larger black circles*, TCN. For convenience, the TC dendritic tree is divided into four quadrants, with the *upper right* quadrant illustrating terminal sites of CT axons, the *lower right* quadrant showing medial lemniscal inputs, and the *lower right* and *upper left* quadrants showing spinothalamic and thalamic reticular axonal terminals, respectively. The *shaded circle* separates the dendrites proximal to the soma from the distal dendrites. **C** Synaptic inputs to TC in the VLp, in which the main subthalamic input is from deep cerebellar nuclei. **D** Synaptic inputs to TC in the oral/medial ventral lateral nuclei (VLo/VLm) with subthalamic input from globus pallidus and substantia nigra. The quadrantic display of the inputs used in the diagram in **B** is also used in **C** and **D**. See text for published references on which this tentative, stylized representation of thalamocortical connections is based

dritic segments proximal to and distant from the soma are arbitrarily demarcated by the shaded ring. The lower quadrants of Fig. 15B show the synaptic organization of the spinothalamic and medial lemniscal inputs, respectively, and the upper quadrants illustrate the synaptic connections of axons originating from the somatotopically matching cerebral cortex and thalamic reticular nucleus. This quadrantic separation of the terminations of the four inputs to each thalamocortical neuron is used for convenience in the figure, although medial lemniscal and spinothalamic synaptic terminals are probably located on different dendritic segments (Ralston and Ralston 1993, 1994a).

The synaptic terminations in VPLc of medial lemniscal fibers are quite complex (Fig. 3.15A,B). Although functionally heterogeneous, with separate neuron populations transmitting mechanoreceptive information from cutaneous (Meissner, Merkel, Pacinian afferents), muscle, joint, and other deep somatic tissues (Talbot et al. 1968; Darian-Smith 1984), medial lemniscal axon terminations are morphologically uniform and, as was seen earlier, are distributed throughout VPLc. Only about 15% of (excitatory) medial lemniscal boutons terminate separately on the dendrites of thalamocortical neurons. The remaining boutons make this contact in association with a third neuronal element, which is a dendritic extension of a local circuit neuron in VPLc and is recognized as such by its ultimate continuity with the soma of the interneuron or by the presence of ribosomes in its cytoplasm, which occur only in dendrites (Ralston 1971; Ohara et al. 1989). The special feature of the interneuronal dendritic elements associated with the medial lemniscal synapse on the dendrites of thalamocortical neurons is that they contain flat or pleomorphic synaptic vesicles. The medial lemiscal bouton filled with round vesicles, the pleomorphic vesicle-filled interneuronal dendrite or "presynaptic dendrite" (for a review, see Ralston and Ralston 1993), and the postsynaptic thalamocortical dendrite together constitute a synaptic triad (Ohara et al. 1989), with the medial lemniscal bouton synapsing on both the presynaptic dendrite and thalamocortical dendrite, and the presynaptic dendrite also synapsing on the thalamocortical dendrite. When additional presynaptic dendrites are incorporated into this complex, and the whole is somewhat separated from adjacent neurons by an astrocytic capsule, it has long been described as a synaptic glomerulus (Jones 1985; Ralston and Ralston 1993, 1994a). Finally, it has been established that the thalamic interneurons in VPLc are GABAergic (Ohara et al. 1989; Ralston and Ralston 1994b) and hence inhibitory to the neurons on which they synapse. Medial lemniscal boutons are distributed mainly on proximal dendrites of thalamocortical neurons.

The synaptic terminals of spinothalamic axons are rather differently organized. Again, these fiber populations are functionally heterogeneous, with about 85% originating from laminae I and II of the spinal dorsal horn and transmitting information from thermoreceptive and nociceptive afferents, and 15% projecting from the deep laminae of the dorsal horn and transmitting information from cutaneous and other somatic mechanoreceptive afferents (Willis and Coggeshall 1991). Unlike medial lemniscal afferents, which terminate entirely within VPLc, spinothalamic neuron terminal arbors are more patchy in their distribution within that nucleus and, in addition, project to other thalamic nuclei, including the oral pulvinar, VPLo, VPI, zona incerta, Cl, and MD (Apkarian and Hodge 1989a–c). Contrasting sharply with the synaptic terminations of medial lemniscal afferents, only about 15% of the boutons of spinothalamic axons terminate on thalamic interneurons; it is possibly these axons which, like medial lemniscal afferents, transmit somatic mechanoreceptive information (Ralston and Ralston 1993). The majority of spinothalamic axon terminals in VPLc form simple axodendritic synapses on thalamocortical neurons, without associated GABAergic interneurons, and hence are not subject to modulation through these neurons. Typically, a whole string of these synapses may be visualized along both the proximal and distal segments of a particular dendrite of a thalamocortical neuron. In addition, it seems that the synapses of medial lemniscal and spinothalamic terminals are usually segregated and not found on the same dendritic segment (Ralston and Ralston 1994a). The traditional view that medial lemniscal and spinothalamic axon terminals do not converge on the same thalamocortical neurons may be correct, but there is some morphological and electrophysiological evidence suggesting that this convergence occurs in the monkey (Ralston and Ralston 1994a,b; Chung et al. 1986). Studies using the new differentiable antergrade labels are needed.

Figure 3.15C,D illustrates current views of the synaptic inputs to thalamic "relay" cells (TC) in nuclei with inputs from the cerebellum and the basal ganglia.

Two populations of corticothalamic axons projecting to VPLc in the macaque were described above, which are characterized by rather different terminal arborizations in that nucleus (A. Tan et al., in preparation). However, only one class of corticothalamic synaptic bouton, packed with vesicles but containing few intracellular organelles, has been identified in electron microscope studies of the macaque VPLc (Williamson and Ralston 1993). These were found to terminate either on presynaptic dendrites of interneurons or directly on the more distal segments of thalamocortical dendrites.

The axons of thalamic reticular neurons are the fourth source of synaptic terminals located on the distal segments of dendrites of thalamocortical neurons in the macaque. These neurons are GABAergic and, as expected, their large flat boutons contain flattened or pleomorphic vesicles. Typically these terminals are not found within thalamic glomeruli and do not form synapses on the other GABAergic element in VPLc, namely the presynaptic dendrites of interneurons (Ohara et al. 1989).

3.6 The Pulvinar, Visually Directed Hand Movements, and Object Recognition

A quite extensive thalamocortical complex which seems to have a special role in the execution of visually directed hand movements is that centered on the pulvinar and lateral posterior thalamic nuclei and their cortical connections. Although a small pulvinar nucleus is identifiable in the cat and tree-shrew (for a review, see Jones 1985), it evolves into a quite large nuclear mass in the primate, roughly in parallel with the ballooning of the posterior parietal and temporal "association" cortex in the prosimians, New and Old World monkeys, apes, and man (Ogren 1993) and the development of the use of the hand in these different primates. As seen in the macaque (Fig. 3.2), the Lgn, the medial geniculate, and the ventral posterior (VPLc) nuclei, with inputs from the retina, cochlea, somatic proprioceptors, and vestibular apparatus and from tactile receptors, are spatially embedded in the mass of the pulvinar. This topographic association in some way reflects the fact that the sensory information processed by each of these neuron populations is integrated to provide a precise, continuously upgraded representation of the physical space in which the subject moves and grasps, handles, and identifies objects within reach. Current evidence suggests that the pulvinar and its connections with the parietal and temporal cortex encompass those neuron populations which have a central role in this complex integration of information about the "world out there" (Acuna et al. 1990; Robinson 1993; Robinson and Peterson 1992).

3.6.1 Topography and Cytoarchitecture

The pulvinar constitutes the most posterior part of the the macaque's thalamus (Fig. 3.2). Most of its boundaries are defined by their obvious topography, but identifying the boundary between the oral pulvinar and the dorsorostral lateral posterior nucleus (LP) is somewhat arbitrary, because of their similar cytoarchitecture. Similarly, because of the relative uniformity of its cytoarchitecture, as seen in serial cresyl violet sections, Olszewski (1952) and most subsequent investigators (e.g., Jones 1985; Darian-Smith et al. 1990a) have relied considerably on topographic features for subdividing the pulvinar nuclear mass into four subnuclei, the oral, medial, lateral, and inferior pulvinar. Coronal sections selected to illustrate these subdivisions are shown in Fig. 3.1; serial stacks of such maps were used to construct the three-dimensional maps of the posterior thalamus in Figs. 3.2–3.4. Olszewski (1952) and Jones (1985) describe their cytoarchitecture; Figs. 3.2 and 3.3 illustrate their topography. The oral pulvinar is located between VPLc laterally and CnMd medially, merging caudally with the medial and lateral pulvinar. The dorsomedial medial pulvinar forms that surface of the pulvinar and fuses ventrally with the rostral midbrain; its poorly defined lateral boundary abuts against the lateral pulvinar, which, with a lighter visual texture in Nissl sections that results from clusters of small neuron somas separated by the mediolateral passage of fibers, forms the dorsolateral pulvinar. Finally, the inferior pulvinar is ventrally located between the medial geniculate nucleus (GM) and the Lgn (GLd); it is densely staining and separated from the rest of the pulvinar by a transverse fiber bundle, the brachium of the superior colliculus.

While Olszewski's anatomical subdivisions, defined mainly by topographic landmarks, provides a useful reference framework for pulvinar space, they have limited functional relevance, as Olszewski (1952) anticipated. However, with improved definition of (a) pulvinar connectivity, (b) single neuron and population response characteristics, (c) chemoarchitecture using the new histochemical neuron labels, such as calbindin, parvalbumin, and cytochrome oxidase (Cusick et al. 1993), (d) the behavioral deficits resulting from focal lesions of the pulvinar and parietal and temporal cortex, and (e) comparative studies in different primate species, functionally relevant subdivisions of the pulvinar are now being sought. Since better and more systematic mapping of the pulvinar is still incomplete, in the following paragraphs we continue to use

Olszewski's description simply as a reference framework in pulvinar space in the macaque.

3.6.1.1 Connections

The pulvinar complex in the macaque differs from other thalamic nuclei that we have already considered to have a sensorimotor role in that its most extensive connections are cortical. These cortical connections are with most of the parietal and temporal cortex (Clark and Northfield 1937; Baleydier and Maugiere 1987; Balydier and Morel 1992; Pons and Kaas 1985; Darian-Smith et al. 1990a; Cusick and Gould 1990; Hardy and Lynch 1992; Robinson and Peterson 1992; Yeterian and Pandya 1991), as well as with the striate and peristriate cortex (Fig. 3.2), and are considered more fully later. However, the pulvinar also has important visual inputs from the retina (Itaya and Van Hoessen 1983; Nakagawa and Tanaka 1984; Cowey et al. 1994) and superior colliculus (Benevento and Fallon 1975; Rodieck and Watanabe 1993) and a somatic spinothalamic input. Each of these has restricted, nonoverlapping terminations within the ventral and anterior pulvinar. In the macaque, the retinal input to the pulvinar includes projections of the three main ganglion cells, namely the Pα cells (or A or M cells), the Pβ cells (or B or P cells), which constitute about 80% of ganglion cells and are thought to mediate color vision, and Pγ cells (Cowey et al. 1994). These fibers terminate in the middle and medial part of the inferior pulvinar (PIc; Cusick et al. 1993) and possibly the medial pulvinar. How this retinal projection matches with the cortically dependent retinotopic maps in both the inferior and medial pulvinar is not known, except that the direct retinal projection is more limited in its distribution. Pulvinar connections with the superior colliculus, which itself has a retinal input, are more extensive and include retinotopic projections to zones which straddle the inferior, medial, and lateral pulvinar.

The most direct pathway which transmits somatic proprioceptive and tactile information to the pulvinar is the spinothalamic projection, some of whose fibers terminate in the oral pulvinar (Berkley 1980, 1983; Apkarian and Hodge 1989a–c). It must be emphasized that, like the visual input, the spinothalamic input to the pulvinar is quite small relative to that coming from cortical areas 3a, 3b, 1, and 2, which also signals information about the body parts and their movement relative to each other and to the external world. Nonetheless, information that is relayed from retinal and somatosensory receptor neuron populations and is still untransformed by processing within sensory cortex may provide key information concerning the immediate state of the external world.

As with thalamocortical projections to other areas of cerebral cortex, each small zone of cortex receives input from several thalamic nuclei. Figure 3.11 illustrates the fractional projections from these different nuclei to the posterior parietal areas 5a, 5b, 7a, and 7b, the major projections originating from adjacent areas of LP and the oral and medial pulvinar. Reciprocal thalamic projections to the inferior temporal cortex originate mainly from the lateral, medial, and inferior pulvinar (Webster et al. 1994); each of these pulvinar zones is the target of subthalamic visual inputs.

Rockland (1994) has recently differentiated two morphologically distinct

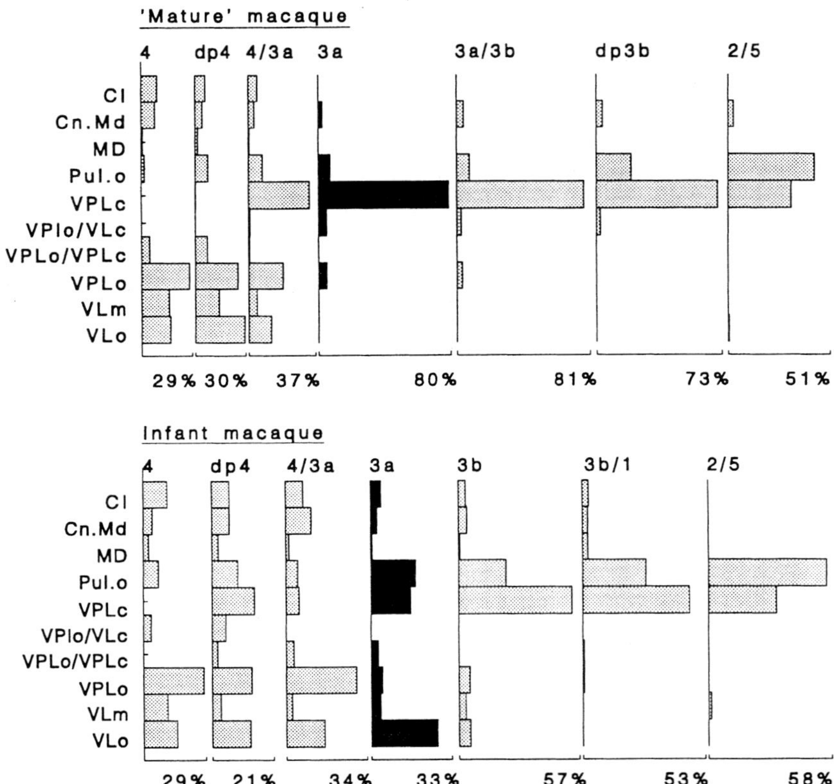

Fig. 3.16. Relative weightings of projections from different thalamic nuclei to cortical areas 3a (*black*) and adjacent areas 4, 3b, and 2/5 in the mature (*top*) and newborn (*bottom*) macaque. Projection from each thalamic nucleus is expressed as a percentage of the total neuron sample projecting to that cortical target. Comparing the data from mature and infant monkeys illustrates the abrupt shift in the nuclear origins of thalamic projections to cortical areas 4, 3a, 3b, 1, 2, and 5 in both mature and infant macaques and the more diffuse and divergent thalamocortical projections in infants (Darian-Smith and Darian-Smith 1993). *Cl*, central lateral nucleus; *Cn.Md*, medial central nucleus; *MD*, medial dorsal nucleus; *Pul.o*, oral pulvinar; *VPLc*, caudal lateral ventral posterior nucleus; *VPI*, inferior ventral posterior nucleus; *VLc*, caudal ventral lateral nucleus; *VPLo*, oral lateral ventral posterior nucleus; *VLm*, medial ventral lateral nucleus; *VLo*, oral ventral lateral nucleus

corticothalamic neuron types projecting from the temporal cortex to the medial and lateral pulvinar, which have many of the features of corticothalamic neurons originating from the precentral sensorimotor cortex (see Figs. 3.14, 3.15). Both of these corticopulvinar neuron types have their pyramidal cell bodies in lamina V, and the terminal branching of the axons of each cell type are dense in the environs of the thalamocortical neurons that are reciprocally related to them. It is not yet known, however, whether these corticothalamic terminals synapse directly on the thalamic relay cells. The commoner corticopulvinar neuron type has a rather fine, fairly straight axon directed toward the soma distribution of the reciprocal thalamocortical cells. However, like its counterpart among corticothalamic axons directed to other sensorimotor thalamic nuclei (Fig. 3.14), this axon gives off along its whole extent many single, slender branches 5–10 μm long which end in single boutons. Such extensive distributions of synapses through much of the thalamic nucleus, and not just in the proximity of the somas of reciprocal thalamocortical cells, imply that these corticothalamic neurons have a diffuse synaptic action within the targeted thalamic nucleus. In contrast, the less common, coarse corticothalamic axons terminate as a cluster of synaptic terminals in proximity with the reciprocal thalamocortical neurons; their action is presumably quite localized.

3.7 Comment

Substantial progress has been made in studies of the functional anatomy of the primate thalamus since the publication of Jones' (1985) monumental review a decade ago. Tracer studies have mostly reinforced the early proposal of Kievit and Kuypers (1977) that implied that there is considerable "territorial" convergence of input from two or more thalamic nuclei to each localized zone of cortex and that each thalamic nucleus, specified by its cytoarchitecture and subthalamic input, projects to a quite extensive area of sensorimotor cortex, which overlaps with the projections of adjacent thalamic nuclei (Fig. 3.16). Thus the thalamocortical projections ensure a highly ordered redistribution of sensorimotor information from the spinal cord, cerebellum, vestibular nuclei, globus pallidus, and substantia nigra to each localized zone of cortex. Each cortical zone will therefore have a unique subthalamic input, specified by the mix of these various inputs.

With increasing awareness among investigators of the extensive areas of cortex that are active during the execution of even rather simple voluntary manual tasks and of the complexities of the thalamocortical projections outlined above, two somewhat neglected aspects of thalamocortical connectivity are now being examined with renewed interest, using the new tracing and intracellular labeling techniques. First, the ubiquitous corticothalamic projections in the primate can no longer be reliably considered to reciprocate the thalamocortical projections with which they are spatially associated. Corticothalamic subpopulations with different connectivity have now been identified and need further elucidation of their microstructure and function. Second, the pulvinar complex, much expanded in the primate brain and with extensive, complex connections with the parietal cortex, seems to have a special relevance to visually guided manual behavior. Unlike the sensory "relay" nuclei, VPLc and the lateral geniculate nucleus (GLd), its major inputs are from the cerebral cortex, and its subthalamic input is quite limited.

4 Sensorimotor Cortex

4.1 Introduction

In the preceding chapters, the focus has been on the functional anatomy of those thalamic and cortical neuron populations which have a direct role in transmitting to and from the sensorimotor cortex information that is important in the execution of manual tasks such as reaching for, identifying, and retrieving nearby objects (see Figs. 1.1, 1.4) or tracking with the finger the position in visual space of a moving object. Somatosensory and visual information together define the world in which the hand and fingers are moved, the location and physical features of the target object, and the spatial relations of the fingers to this object. The corticospinal neuron populations together transmit to the spinal cord neuron populations information which will shape motoneuron activity and the coordinated hand and finger movements appropriate to the behavioral goal. The effectiveness and sophistication of the finger/hand movements in achieving this behavioral goal ultimately depend on the *intervening* neuronal processing that occurs in the cerebral cortex and its subcortical connections. In this chapter, we briefly examine these cortical connections.

When considered as a whole, the task of reaching and retrieving, as with any sensorimotor action, is sequential. However, several recent investigators (McClelland et al. 1986; McClelland 1989; Arbib and Hoff 1994; Jeannerod 1994a,b; Goldman-Rakic 1987b, 1988a,b) have emphasized that this sequencing does not necessarily imply a succession of behavioral (or neuronal) ministeps, each of which must be completed before the next starts. Rather, there will be overlap or coincidence of many of these steps. For example, the continuing assessment of the extrapersonal and intrapersonal space and of the relative positions of the target object and fingers occurs coincidentally with reaching out and grasping an object with the fingers. In fact, we have seen that the complexity of the simplest manual task dictates that several sequential control systems operate in parallel with each other and are coordinated in their actions to achieve the final behavioral goal. The cumulated evidence concerning its functional anatomy and connections, the behavioral deficits resulting from localized lesions, single neuron responses in different areas, and recently positron emission tomography (PET) studies of regional activity during the execution of different manual tasks all point to the cerebral cortex having a major role in the coordination of these multiple, coactive control systems that operate in parallel in any voluntary action. This idea is hardly novel and has been embedded in the literature of "association" cortex for more than a century. However, during the last decade our thinking about the role of the cerebral cortex in voluntary action has been shaken up by the following insights:

- The recognition that neuron populations in a number of quite diverse areas of cortex are coactive during the performance of apparently simple manual tasks and that the responses of *single* neurons in these separate populations may at times look remarkably similar (Darian-Smith et al. 1985; Kalaska et al. 1990; Fetz 1992; Cohen et al. 1994)
- The identification of a bewildering assembly of connections between cortical areas in which these coactive cortical neuron populations are located (Asanuma et al. 1985; Andersen et al. 1990b, 1993a,b; Goldman-Rakic 1987, 1994)
- The appreciation that the neuronal representation of any component of this sensorimotor behavior may be spread across or distributed in several discrete subsystems operating in parallel, i.e., parallel distributed processing is the norm in these key pathways (McClelland et al. 1986)

The difficulty lies in making sense of the more than 300 connections that have now been described in the macaque's cerebral cortex (Merigan and Maunsell 1993). A special difficulty arises from the fact that this present description is mainly of the macrostructure of cortical *projections* and is quite limited in defining the microstructure of *connections*, i.e., synaptic organization. This imbalance in the available description of cortical projections and connections greatly limits recent attempts at modeling the neuronal circuitry mediating voluntary action, as is apparent in the following account.

In this chapter, only selected aspects of the functional anatomy of the connections of the macaque's cerebral cortex as they relate to manual dexterity are reviewed. First we examine the connections of the visual and somatosensory cortical areas which provide the necessary flowpaths for distributing this information to specific neuron populations in the posterior parietal and temporal cortex. These connections provide a framework for channeling visual and somatosensory (and vestibular) information about extrapersonal and intrapersonal space to the posterior parietal cortex and information about the visual and tangible features of the objects handled to the inferior temporal cortex. We then examine those cortical connections between the posterior parietal, temporal, cingulate, and prefrontal cortex which might further contribute to the intelligent handling of nearby objects. The focus is on ipsilateral cortical connections. Finally, we focus on the cortical connections of the several areas in frontal, cingulate, parietal, and insular cortex from which the different corticospinal projections originate. These pathways illustrate the potential for neuron populations in quite diverse areas of cortex to modulate the corticospinal signals transmitted to the cervical spinal cord and ultimately to the motoneuron populations that determine hand and finger movements.

Before examining the issues outlined above, some PET studies of the changing patterns of regional neuronal activity in the cerebral cortex of human subjects during the execution of different manual tasks are considered. With PET techniques, regional neuronal activity is indirectly assessed by measuring changes in regional cerebral bloodflow or metabolic activity (Deiber et al. 1991; Raichle et al. 1994). While this noninvasive technique tells us little about the timing of

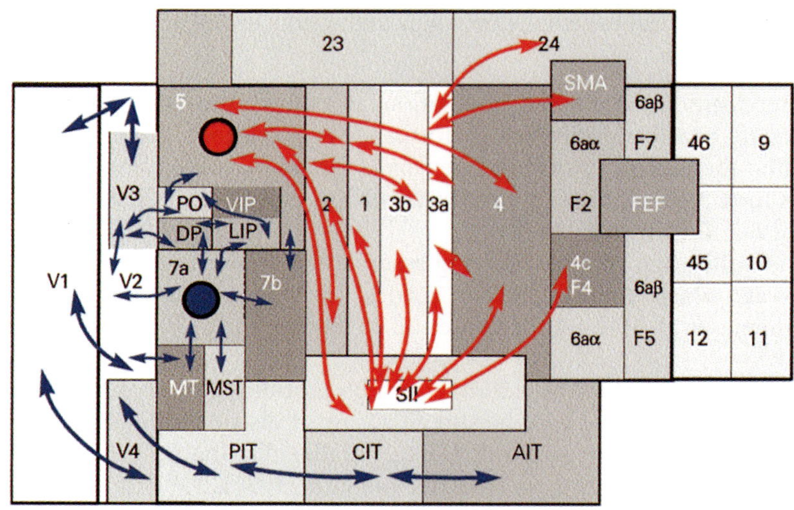
23
24
SMA
6aβ
5
6aα
F7
46
9
V3
PO
VIP
DP
LIP
2
1
3b
3a
4
F2
FEF
4c
F4
45
10
V1
V2
7a
7b
6aβ
6aα
F5
12
11
MT
MST
SII
V4
PIT
CIT
AIT
A

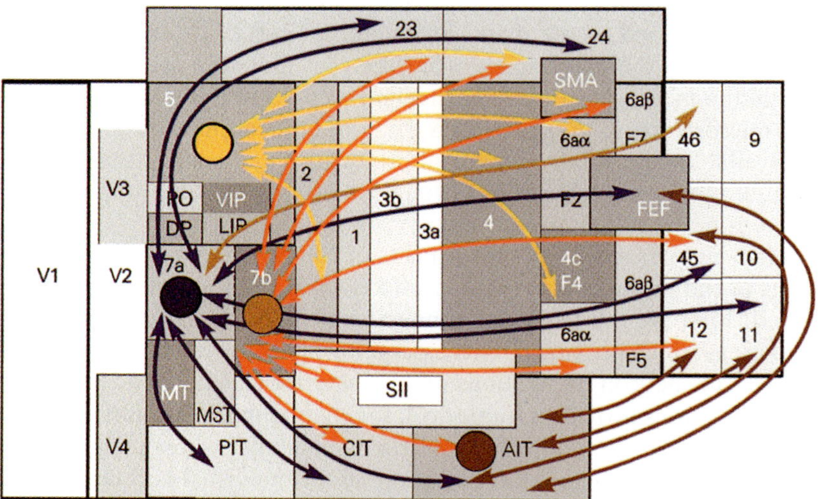
23
24
SMA
6aβ
5
6aα
F7
46
9
V3
PO
VIP
DP
LIP
2
3b
1
3a
4
F2
FEF
4c
F4
45
10
V1
V2
7a
7b
6aβ
6aα
12
11
F5
SII
MT
MST
V4
PIT
CIT
AIT
B

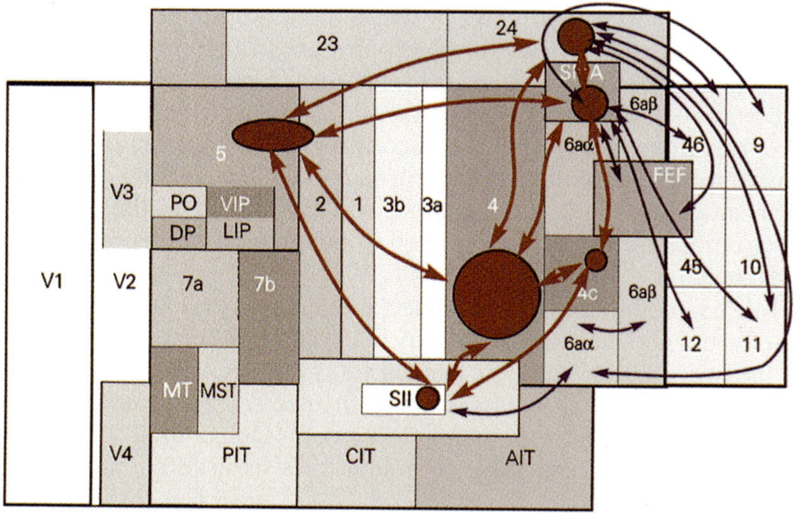
23
24
SMA
6aβ
5
6aα
46
9
V3
PO
VIP
DP
LIP
2
1
3b
3a
4
FEF
45
10
V1
V2
7a
7b
4c
6aβ
6aα
12
11
MT
MST
SII
V4
PIT
CIT
AIT
C

neural events, it does provide important information about *where* increases in neuronal activity occur in the human cerebral cortex during the preparatory and executive phases of complex manual tasks.

With repetitive finger flexion–extension, an increase in neuronal activity occurs in the region of hand/finger representation in the contralateral sensorimotor areas 4 and adjacent postcentral cortex (Roland et al. 1980a,b; Colebatch et al. 1991; Seitz and Roland 1992; Shibasaki et al. 1993; Remy et al. 1994), but has not been detected in SII. However, with a slight increase in the complexity of the finger movement, such as flexion–extension of the middle finger only or a sequencing of finger movements requiring selective attention or some planning, neuronal activity increases bilaterally in the supplementary motor area (SMA) and in some subjects also in the contralateral premotor cortex (Shibasaki et al. 1993). In the case of manual tasks that include the tactile identification of the shape of an object held in the fingers, Seitz et al. (1991) report a further extension of regional neuronal activity to include the contralateral superior parietal lobule (area 5) and small increases in the cingulate cortex and parts of the prefrontal cortex bilaterally. As expected from these observations, the regional cortical activitation (estimated from regional bloodflow measurements) associated with visually guided manual tracking tasks is even more extensive (Deiber et al. 1991; Grafton et al. 1992, 1993), including, in addition to the now bilateral pericentral sensorimotor cortex in the region of the hand representation, the bilateral dorsal and mesial parietal cortex, the SMA, and the cingulate cortex. The precision in the PET mapping of active cortical regions has greatly increased in recent studies, but is of course still limited. Finally, a number of PET studies have now shown that the various patterns of regional activity associated with the execution of different motor tasks can be substantially modified when the task requires complex preplanning, different levels of selective attention, motor learning, or memory recall (Colebatch et al. 1991; Deiber et al. 1991; Decety et al. 1992; Raichle et al. 1994; Wessel et al. 1994). Specifically with manual tasks, these activity maps of the cortex may be greatly modified following localized forebrain lesions, such as an intracapsular lesion that interrupts the corticospinal projections (Weiller et al. 1992a,b).

These PET studies provide a backdrop for the following comments on the functional anatomy of cortical connections relevant to the performance of voluntary manual tasks. They highlight the fact that many areas of cerebral cortex are coactive and have a role in this behavior.

◀ **Fig. 4.1.** Corticocortical connections of the neocortex in the macaque. Connections are plotted onto a stylized unfolded neocortex in which the different cytoarchitectonic (or functionally distinct areas) are shown (see Fig. 2.1 for map). Most, but not all connections shown are reciprocal. **A** visual (*blue*) and somatosensory projections (*red*) to posterior parietal and inferior temporal cortex. Areas 5, 7a, and 7b have especially important convergent and divergent connections. **B** connections of parietotemporal and frontocingulate cortical areas. **C** interconnections of the different corticospinal neuron populations in the frontal, cingulate, parietal, and insular cortex and their connections with prefrontal cortex. *FEF,* frontal eye fields; *SMA,* supplementary motor area

4.2 Distribution of Visual Information in the Parietotemporal Cortex

About 90% of retinal ganglion cells relay in the lateral geniculate nucleus (LGN) of the macaque (Leventhal et al. 1981) about 80% of these (P cells) constitute a morphologically and functionally distinct population whose axons terminate in the parvocellular layers of the LGN, while the axons of another 10% of morphologically distinctive cells (M cells) terminate in its magnocellular layers. In turn, the projections from the magnocellular and parvocellular layers of the LGN remain segregated and terminate in different layers in the striate cortex, the projection from the magnocellular geniculate layers terminating in layer 4Cα and that from the parvocellular layers in cortical layers 4A and 4Cβ. The main functional differences in cells in the M and P subcortical pathways is that only those in the P pathway are color sensitive, and these cells respond in a more sustained fashion to stepwise increases in illumination, but are less responsive to changes in contrast than cells in the M pathway. In other respects the two populations are functionally quite similar.

On tracing the cortical projections of striate cortex, two separate processing streams, one into the posterior parietal cortex and the other into the inferior temporal cortex, have been identified by a number of investigators. The blue arrows in Fig. 4.1A illustrate those visual processing streams which extend forward from the striate cortex, and Fig. 4.1B shows some of the parietal connections in more detail (for reviews, see Zeki and Shipp 1988; Merigan and Maunsell 1993; Stein 1992; Martin 1992; Lachica et al. 1992; Andersen et al. 1990a,b,

Table 4.1. Abbreviations for cortical fields in macaque monkey

Cortical area	Abbreviation	Key references
V1, V2, V3, V4	Visual cortices	Figure 2.1 and legend; Merigan and Maunsell 1993; Stein 1992
1, 2, 3a, 3b, 4, 4c, 5, 6aα, 6aβ, 7a, 7b, 9, 10, 11, 12, 23, 24, 45, 46	Cytoarchitectonic cortical fields	Brodmann 1989; Vogt and Bogt 1919; Walker 1940; Matelli et al. 1989, 1991
F2, F3, F4, F5, F6, F7		Matelli et al. 1989, 1991
SMA	Supplementary motor area	Penfield and Welch 1951
SII	Somatosenory area II	
FEF	Frontal eye field	
VIP	Ventral intraparietal cortex	
PO	Preoccipital cortex	
DP	Dorsal parietal cortex	
LIP	Lateral intraparietal cortex	Seltzer and Pandya (1989, 1994) for assessment of subdivisions of parietal, occipital, and temporal cortex in macaque
MT	Middle temporal cortex	
MST	Middle superior temporal cortex	
PIT	Posterior inferior temporal cortex	
CIT	Central inferior temporal cortex	
AIT	Anterior inferior temporal cortex	

1993a,b; Blatt et al. 1987). Ungerleider and Mishkin (1982) proposed two cortical processing streams for visual information, one extending through dorsally located prestriate areas (V2, V3, LIP, MT, and, middle superior temporal cortex, MST) to the posterior parietal cortex (mainly area 7a), and the second through the ventral prestriate areas (V4) to the inferior temporal cortex (posterior, middle, and anterior: PIT, MIT, and AIT; see Table 4.1 for a list of abbreviations). The dorsal system was considered to process visuospatial information and the visible motion of objects, and the ventral system to process information concerning the form, pattern, and color of objects. Livingstone and Hubel (1988) and others (Maunsell 1987; Zeki and Shipp 1988) developed the concept that these cortical pathways operate as *functionally segregated* extensions of the subcortical M and P pathways, the parietal pathway mediating behavior associated with the assessment of visual space and the movement of target objects in that space, while the visual neuron populations in the more ventral pathway were considered, as Ungerleider and Mishkin had suggested for the cortical channels, to mediate the identification of color and the form of target objects. Superficially, this model provided an attractive basis for explaining the behavioral deficits resulting from focal lesions in the posterior parietal and inferior temporal cortex in human and macaque subjects, but recent critical studies (reviewed by Merigan and Maunsell 1993) have shown that neither anatomical nor functional segregation of the visual processing pathways in the parietal and temporal cortex is as complete as it was earlier thought to be. Rather, anatomical cross-connections exist, which might mediate cross talk; further, there is substantial, although by no means complete, overlap in the response characteristics of neurons in the two streams. In other words, the two processing streams operate in *parallel*, and some information about the different features of the visual behavior that they mediate is *distributed* in the parallel pathways, rather than being segregated in them, as was proposed by Livingstone and Hubel. It will be recalled that every part of the posterior parietal and inferior temporal cortex, in addition to its cortical connections, has quite dense "reciprocal" connections with specific neuron populations within the pulvinar (see Fig. 3.2), providing a substrate for further interaction between the parietal and temporal visual pathways.

Much of the early processing of visual information within the cerebral cortex occurs within the parietal and temporal areas receiving input from the striate cortex (V1), before being distributed to other more rostral cortical fields. This is so because only cortical areas LIP, 7a, 7b, 7e, and 5 have cortical connections extending beyond their immediately surrounding areas (Fig. 4.1B). These connections are further considered below, with the view to locating those areas of parietal cortex in which visual and somatosensory information may be correlated and used to construct an accurate representation of the external world and of the position and movement of the hand (and other body parts) in that space.

4.3 Flow of Somatosensory Information in the Parietal Cortex in the Macaque

It will be recalled that in the macaque each of the postcentral areas 3a, 3b, 1, and 2, has a separate, somatotopically organized input from the thalamic caudal lateral ventral posterior nucleus (VPLc), as well as some input from the anterior pulvinar (Jones and Friedman 1982; Jones 1985; Cusick et al. 1985; Pons and Kaas 1985; Darian-Smith et al. 1990a). Recording from single neurons in these cortical fields has demonstrated an orderly shift in the receptive field characteristics both mediolaterally along the postcentral gyrus and rostrocaudally across it (Powell and Mountcastle 1959; Mountcastle and Powell 1959; Kaas et al. 1979, 1981; Nelson et al. 1980; Darian-Smith et al. 1982, 1984, 1985; Iwamura and Tanaka 1978; Iwamura 1993). Mediolaterally within cortical areas 3b and 1 is the long recognized somatotopic representation of the body surface, the cutaneous receptive fields of individual neurons often closely resembling those of peripheral mechanoreceptive afferents innervating the same region of skin. Thus in the alert, active monkey in the area of 3b/1 cortex with input from the distal pad of the contralateral index finger or thumb, cells usually have cutaneous receptive fields only a few millimeters in diameter, with response characteristics quite similar to those of rapidly adapting Meissner afferents or slowly adapting Merkel afferents (Talbot et al. 1968; Darian-Smith et al. 1984, 1985; Iwamura and Tanaka 1978; Iwamura 1993). Caudally, in area 2, the receptive fields of individual cells are typically larger, often extending over several fingers, and they may be discontinuous, with responsive zones on the terminal pads of several adjacent fingers. A substantial fraction of these cells are direction selective in their responsiveness to cutaneous stimuli. Rostrally, in area 3a, the response characteristics of individual neurons again change. These cells have been incompletely characterized in the alert, active macaque), but from the limited studies in anesthetized monkeys few of them have well-defined cutaneous receptive fields; their peripheral input is from deep somatic tissues and probably from muscle and tendon mechanoreceptive afferents (Hore et al. 1976). If this is so, it would be expected that cells in area 3a with inputs from any part of the moving arm, hand, and fingers might be active while reaching for and retrieving an object, contrasting with cells in areas 3b and 1, in which only those neurons with cutaneous receptive fields on the digital pads would discharge, and then only during the "tactile phase" of the manual task.

The stream of red arrows in Fig. 4.1A illustrates the transmission of somatosensory information caudally from areas 3a and 3b to areas 1 and 2 and then to area 5. Projections from SII reinforce this stream. Mountcastle et al. (1975) showed that the response characteristics of neurons in area 5, when recorded in the alert, active macaque, are substantially more diverse and more complex that those of cells in area 2. In addition to having more extensive receptive fields than those of area 2 cells, many neurons responded best to active movement of the relevant joint(s) or to stimuli moving tangentially across the skin rather than just indenting the skin. Other neurons in area 5 responded best when the monkey reached for an object of interest, and yet others when the object of interest was actively manipulated. Selectively attending to the manipulated object seemed to enhance the discharge in many of these neurons. These novel findings have subsequently been confirmed and extended (e.g., Kalaska and Hyde 1985). Some neurons in the rostral part of area 7 (area 7b) have somatosensory input rather similar to those in area 5, but especially in the caudal part, in area 7a, most neurons are responsive to light and have quite large visual receptive fields (Robinson and Goldberg 1978; Andersen et al. 1990b, 1992). Many of these cells also respond in an orderly way to eye movements.

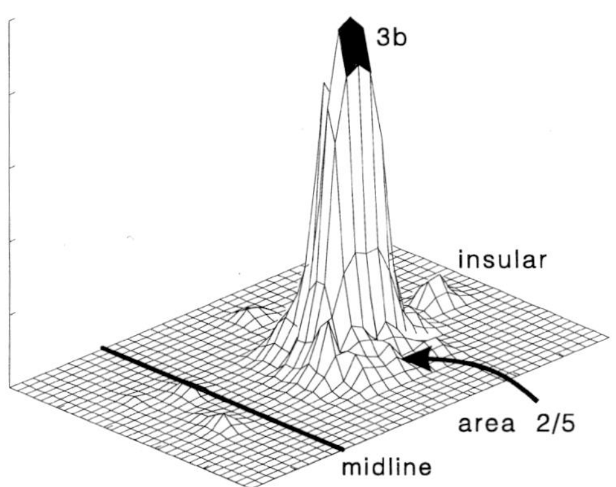

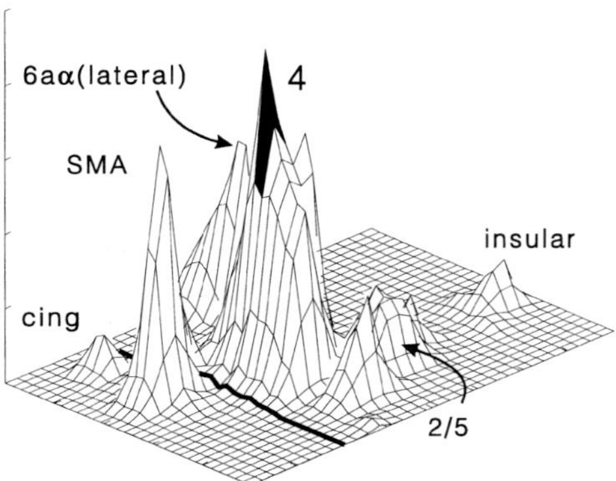

Fig. 4.2. Distributions of local and remote afferent cortical connections of somatosensory and motor cortex in the macaque. The cerebral cortex of the right hemisphere has been unfolded to form a planar surface on which the distribution of retrogradely labeled somas of neurons projecting to a localized zone (less than 1 mm) in areas 3b (*top*) and 4 (*bottom*) are plotted. In somatosensory area 3b, most afferents originate from neighboring cortex, but a few have their somas in the supplementary motor area (*SMA*), cingulate, parietal, or insular cortex. In contrast, in motor cortex (e.g., area 4) convergence of input is the rule, so that the proportion of cortical afferents originating from distinctive remote neuron populations is quite substantial (Darian-Smith et al. 1993)

Figure 4.2 illustrates two of the patterns of cortical afferent connections of areas 3a, 3b, and 4 in the macaque. First, each small cluster of neurons in each of these cortical areas receives input from the immediately surrounding cortex (3–4 mm in radius), probably through intracortical axons. In addition, each such neuron population has connections with particular, more remote cortical areas. The simplest pattern of these connections is seen with areas 3b and 1, which connect only with areas 2/5 and SII. In contrast, the remote cortical connections of area 4 are quite extensive (Fig. 4.2), afferent inputs originating from the anterior cingulate cortex, the SMA, and the postarcuate cortex, insular cortex, and parietal areas 2 and 5; these are reciprocated (Jones et al. 1978; Petrides and Pandya 1984; Cusick et al. 1985; Pearson and Powell 1985; Leichnetz 1986; Pons and Kaas 1985; Ghosh et al. 1987).

4.4 Visual Space, Tangible Space, and Their Neuronal Alignment

Any combination of cortical neuron populations generating the representation of physical space in which the subject moves and performs manual tasks must have the appropriate inputs of visual, tactile, proprioceptive, and vestibular information. Hence, the posterior parietal cortex, including areas 5, 7a, 7b, and LIP, becomes a prime, but not necessarily unique candidate for this complex processing (Stein 1992; Andersen et al. 1990a,b, 1993b; Blatt et al. 1987; Goldman-Rakic 1988a,b; Cavada and Goldman-Rakic 1989a,b, 1993; Jeannerod 1988, 1994a,b; Crammond and Kalaska 1989; Faugier-Grimaud et al. 1985). Figures 1.2, 1.3, 3.2, 3.4, and 4.1, briefly summarize a large body of investigations of its connections in the macaque. Early clinical studies (Critchley 1953; see also Mountcastle et al. 1975) and more recent experimental analysis of the effects of posterior parietal lesions in the macaque on hand function (Faugier-Grimaud et al. 1985; LaMotte and Mountcastle 1979) strongly reinforced the idea that these areas of the posterior parietal cortex have a role in the central representation of the external physical world and the position and movement of the hand, head, limbs, and trunk in that space and, further, that this is a role shared with some other cortical areas.

4.4.1 Which Spatial Parameters Need Central Representation?

The neural processes which relate and combine the many different visual, tactile, and proprioceptive cues to generate the best representation of the external world and the body parts within it are only now beginning to be examined in any systematic way. A retinotopic representation of the visual world and the position and movement of the hand in that context is continuously represented in the striate cortex (V1) and the adjacent extrastriate areas V2, V3, and V4. Tactile information generated by grasping and manipulating an object in the hand is relayed to the postcentral cortical areas 3a, 3b, 1, 2, 5, and SII, as is proprioceptive information, and locates this input within the framework of somatotopic space. However, neither of these representations provides an adequate description of the space around us or of our location within that space, information that is critical for any dextrous use of the hand. The location of the image of an object on the retina changes with movements of the eye, head, and trunk; similarly, the position of the hand and fingers relative to this object in external space changes with the position of the arm and trunk. To be fully represented in the cerebral cortex, the space around us must therefore be defined in terms of several maps, each with its own set of coordinates. Visible space can be related to the retina, to the position of the eyes within the orbit, and to the position of the head and trunk in the external world. Similarly, tactile space can be defined relative to the digits, to the hand, and to the position of the forelimb and shoulder in space. The additional key process in generating a useful neuronal representation of this external world consists in the continuous cross-matching of these multiple, centrally generated visual and somatosensory representations. This is required so that the changing location of the representation of particular objects within each map that results either from object movement in external space or from eye, head, trunk, or limb movements can be differentiated.

One way of generating these multiple cortical representations of the visual and tangible world would be to have a succession of aligned topographic maps combining in various ways inputs from the retina and vestibular apparatus and the proprioceptive inputs from the neck muscles, trunk, shoulder, hand, and digits. A number of problems arise in constructing such complex maps, including the nonlinear representation of space in the known retinotopic and somatotopic maps in the striate and postcentral cortex and the different coordinate systems used for generating each map (Stein 1992). Thus the striate, cortical representation of retinotopic visual space in the region of the fixation point is much greater than that of the peripheral visual field. Likewise, the cortical representation in areas 3b/1 of the digits and hand constitutes a disproportionately large fraction of the total representation of the contralateral body surface. In fact, there is little current evidence for any such multiple representations in the primate parietal cortex. For example, no topographic cortical representation of visual space independent of eye movements has been reported. While no such alignment of these different topographical maps has been demonstrated in the primate cortex, there is evidence of this organization in the tectum of the barn owl (for a review, see Knudsen and Brainard 1995), in which topographic maps of retinotopic visual space and of the location of a sound source in space are aligned to generate a unified representation of the external world relative to the head. In this species, intraorbital eye movements are not a confounding factor, since they are limited to a few degrees. Multiple topographic representations of this type may possibly also occur in the midbrain in the mammal, but these would provide only a partial representation of extrapersonal space, being influenced by both eye and head position and movement.

Another possible way of fully representing physical space in cortical neuron populations, for example in the posterior parietal cortex, is not based on multiple topographic maps of space, but rather on computing visual extraretinal frames of reference from the basic retinal map and combining these with vestibular and proprioceptive maps to generate an effective representation of the external physical world and the position and movement of the body parts in that space (Andersen et al. 1993a,b; Andersen and Brotchie 1992; Stein 1992; Caminiti et al. 1991; Caminiti and Johnson 1992).

Figure 4.3 illustrates a model proposed by Andersen and coworkers (Andersen 1995; Andersen et al. 1993a,b) for generating multiple representations of space which are independent of eye movement, head movement, and body movement and which provide the spatial framework within which accurate reaching and grasping of a visible object can be achieved. Each shaded box represents one or more cortical neuron populations, whose topography is still poorly defined. The first box on the left might represent certain neuron populations in the posterior parietal cortex, although this is not certain. At the single neuron level, visual and somatosensory space is topographically represented in the primary sensory areas, such as V1,V2, V3, and V4 and also areas 3a, 3b, 1, and 2. The receptor sheet is represented topographically, as a retinotopic or a somatotopic map, so that, for example, depending on the location of the cell's receptive field, the visual response will change with changes in the eye, head, or body position. If,

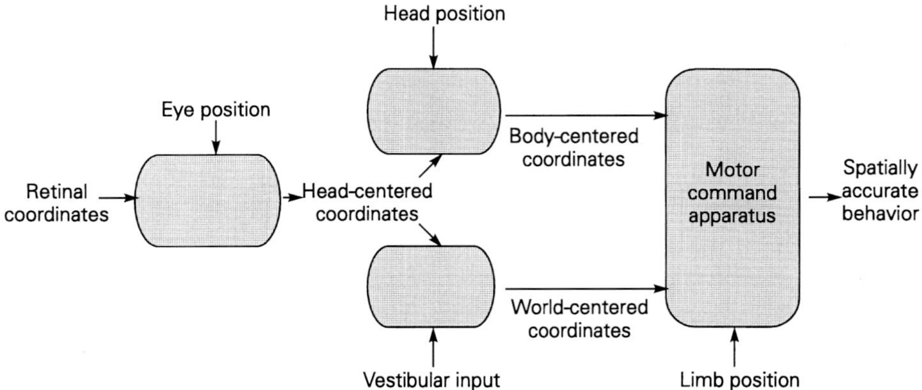

Fig. 4.3. Model proposed by Andersen et al. (1989, 1993a) for generating multiple representations of space which are independent of eye movement, head movement, and body movement and which provide the spatial framework within which accurate reaching and grasping of a visible object can be achieved. Each *shaded box* represents one or more cortical neurom populations. See text for explanation. There is certainly some evidence for the representation in posterior parietal neuron populations of visual space in terms of the different coordinate systems shown in this figure (see Andersen et al. 1993a,b). However, information concerning similar representations of intrapersonal and external space in other neuron populations is still fragmentary

instead of focusing on the single neuron responses, those of cell groups or *populations* are considered, then some neuron populations have been identified in the posterior parietal cortex (Andersen et al. 1989) in which the responses of the constituent cells change with movement of the eyes, but this change is linearly related to the eye movement along the x- and y-coordinates. While the single neuron responses confound information about retinotopic location and eye position, the responses of groups of such cells can accurately define both of these measures. In other words, a representation of visual space is generated that is independent of eye movements; visual space is now related to head position. As shown in Fig. 4.3, the neuronal representation of visual space in terms of head coordinates can be just one step in the generation of other such representations, which are now linearly modulated by vestibular input to generate a new representation of space independent of head movements. A similar cascading of modulating inputs, now from proprioceptors and/or cutaneous inputs, can generate representations of space independent of the position of the eyes relative to the body position or hand position. Whether these transformations occur sequentially, as shown in the diagram, or in parallel is not known.

The experimental data currently available suggests that neuron populations in both the posterior parietal cortex (areas LIP and 7b, with input from area 5; Andersen et al. 1993a,b) and the motor and premotor cortex (areas 4 and 6a; Caminiti et al. 1991; Caminiti and Johnson 1992) may contribute to the computation of the cortical representation of the "world out there." Some single neurons in these cortical populations may be shared in the circuitry that generates the several different representations of space (Andersen et al. 1993a). Such processing would be economical in terms of the total cortical neurons committed to it.

4.5 Connections Between Parietotemporal, Cingulate, and Prefrontal Cortex

Figure 4.1B illustrates the main connections in the macaque between the caudal parietal and temporal cortex and the rostral frontal, prefrontal, and cingulate cortex. Most of the parietal connections are with areas 7a, 7b, 5, and LIP. Three main groups of projections link these cortical areas to the cingulate (areas 23, 24), frontal, prefrontal (SMA, 6a, 12, 45, 46, and frontal eye field, FEF), and temporal cortex (PIT, MIT, and AIT). The latter temporal areas in turn are reciprocally connected to prefrontal areas 11, 12, 45, and the FEF.

An extended description of these projections is not presented here, since quite extensive reviews have been recently published by the groups actively examining them (Goldman-Rakic 1987a,b; Goldman-Rakic et al. 1992; Cavada and Goldman-Rakic 1989a,b, 1993; Andersen et al. 1990, 1993a,b; Merigan and Maunsell 1993; Stein 1992; Webster et al. 1994; Morecraft et al. 1992; Vogt and Pandya 1987; Vogt et al. 1987; Zarzecki 1991). Table 4.2 supplements Fig. 4.1 and lists the physiological and connectional characteristics of the main neuron populations in areas 5, 7a, and 7b.

The most immediate problem in understanding the role of the numerous corticortical connections is that they must be defined more fully. Many corticocortical projections have now been described, but few have been specified in terms of their actual synaptic connections. 'Reciprocal' cortical connections must also be specified at the synaptic level. Physiological recording of single neuron responses of these cells has certainly advanced the characterization of some of these synaptic links. However, visualization of labeled pre- and postsynaptic elements (by light or electron microscopy) and development of a reliable description of the distributions of known synaptic terminals on the soma/dendritic tree of individual cortical neurons are still a technical tour de force and have not been used systematically (see, however, Liu et al. 1995a,b). Using laminar patterns of cortical connectivity (Felleman and Van Essen 1991; Andersen et al. 1990a), in which the laminar termination of axon terminals and

Table 4.2. Comparison of cortical areas 5 and 7 of the macaque. (Adapted from Andersen et al. 1990a)

Features	Area 5	Area 7b	Area 7a
Physiological	Somatosensory, many cells most responsive to active joint and manipulative movements	Somatosensory, similar but not identical to area 5 cells	Visual, with large receptive fields, cells responsive to eye movement, saccades
Connections			
Somatosensory	Areas 3a, 3b, 1, 2, SII	Area 5, insular c.	
Extrastriate	–	PO, MST, STP, IT	PO, DP, LIP, MST, FST, IT
Inferior temporal	–		TFI, TFm, prosubiculum
Cingulate		24, 23	24, 23
Pulvinar	Oral pulvinar	Medial pulvinar	Medial pulvinar

the laminar locations of the somas of efferent neurons are used to assess the likely dominant direction of information flow, is an indirect, uncertain procedure.

4.6 Cortical Connections of Corticospinal Neuron Populations

Figure 4.1C illustrates the locations of the different corticospinal neuron populations which terminate in the cervical spinal cord and which regulate contralateral hand movements (red circles and ellipses). The arrows indicate the reciprocal cortical connections between these corticospinal populations as well as connections with the prefrontal cortex and the FEF. The interconnections between the different corticospinal neuron populations provide the circuitry for ensuring that their outputs to the spinal cord are continuously coordinated. It is also apparent from the three maps (Fig. 4.1A–C) that each corticospinal neuron population has its own unique pattern of cortical connections and that each could relay unique information to the cervical motoneuron populations. Thus each neuron population in the contralateral cortex can, in principle, gain continuous access through these oligosynaptic pathways to cervical motoneuron populations and, in fact, influence hand usage. Such a model of cortical regulatory action is, of course, too generalized to be of much use, but does warrant further examination.

5 Summary and Comments

This review has been mainly focused on the functional anatomy of those thalamic and cortical neuron populations in the macaque brain which are now recognized to have a significant role in the performance of skilled manual tasks. Over the last decade there have been substantial developments in neuroanatomical tracer and immunohistochemical labeling techniques, in the noninvasive visualizing positron emission tomography (PET) and functional magnetic resonance imaging (MRI) procedures, and also a greatly increased number of studies based on the recording of single cortical neuron responses in the macaque during the execution of complex manual and other motor tasks. We are in the midst of a major transformation of our understanding of the neural processes mediating complex sensorimotor behavior. This is both exciting, but also confusing, since few research groups can keep abreast of the important factual and technical advances that are occurring along a broad front. It is for this reason that in this review only oblique reference is made to the burgeoning field of immunohistochemical labeling of thalamic and cortical neurons and their transmitters. It is hoped, however, that this review provides a structural framework for cellular neurobiologists interested in these aspects of thalamocortical and corticospinal organization.

The most important aspect of this ballooning of information concerning the structural organization of the primate sensorimotor thalamus and cortex is that the previous, rather simplistic modeling of thalamocortical, intracortical, and corticospinal neuronal processing is now being seriously questioned, and new, more convincing models are being developed. The recognition that parallel distributed processing (McClelland 1989; McClelland et al. 1986) is a strategem used in every part of the brain for rapidly transferring large amounts of information between coactive neuron population has been important; in this review, we have explored to a degree the implications of this form of neuronal processing for sensorimotor behavior. Most of the recent progress concerning the structural organization of the relevant neuron populations has stemmed from using the modern tracer techniques, with an emphasis on *macrostructure*; these methods tell us a great deal about *projections*, and much less about *connections*. As a consequence, considerable advances in defining the cortical origins and spinal target zones of multiple corticospinal neuron populations have occurred, including the recent identification of direct corticospinal projections from the anterior cingulate cortex (e.g., Hutchins et al. 1988). Similarly, the complexities of the various thalamocortical projections are now better defined than previously (Darian-Smith et al. 1990a,b), as are the more than 300 projections linking different areas of the macaque cerebral cortex (Merigan and Maunsell 1993).

Contrasting with the great advances in identifying these many projections is the much slower progress made concerning their *microstructure*. Specifying the all-important structural organization of the afferent and efferent synaptic connections of each of these many pathways is a singularly demanding and time-consuming task with the currently available techniques. As a consequence, we know relatively little concerning the microstructural organization of most of the neuron populations considered in this review. Until this severe limitation in our knowledge of the important sensorimotor neuron populations is eliminated, the important step of specifying the relevant neuronal circuitry mediating corticospinal, thalamocortical, cortothalamic, and cortical interactions will be delayed. The connectivity of corticospinal neuron populations within the spinal cord segments seems to be the most amenable to analysis with currently developed tracer techniques and will doubtless be the subject of systematic investigations during the coming decade. Elucidating this circuitry constitutes an important step in the analysis of the role of the multiple corticospinal neuron populations as determinants of manual dexterity. A second challenging problem is the connectivity of the various corticothalamic neuron populations in the macaque's sensorimotor cortex. Until the synaptic organization of this very large system is better elucidated, we shall continue in our ignorance of its functional role (Jones 1985).

While identifying the structural framework that can sustain parallel distributed processing in the various sensorimotor pathways is an important advance, this simply sets the stage for tackling the difficult problem of *how* this processing occurs. Parallel distributed processing is intrinsically *processing by neuron populations*, which is difficult to analyze in the active monkey with the techniques we currently use. Georgopoulos et al. (1983; Georgopoulos 1986, 1991) made the important step of demonstrating the vectorial representation of the direction of arm movement in reaching for a target object in the responses of the *coactive population of neurons* in the macaque's precentral motor cortex. More recently, Andersen et al. (1990a,b, 1992, 1993b) and others have assessed neuron populations responses to complex stimuli in different visually responsive areas of the posterior parietal cortex, using small samples of cells and more manageable analytic procedures. These methods promise to be applicable to the analysis of the response characteristics of other posterior parietal neuron populations.

Finally, a quite separate issue is raised. Manual dexterity is uniquely primate in character (Napier 1980) and cannot be experimentally examined in nonprimate species. Further, the structural and functional organization of the primate sensorimotor thalamus and cortex is so profoundly different from that of nonprimate mammals (such as the cat) that interspecies comparisons of different neuron populations is both difficult and uncertain. In the coming decade, the noninvasive MRI, PET, and electromagnetic stimulation procedures will probably replace invasive procedures in some studies. However, we see the experimental use of macaques as continuing to be critically important in studies of sensorimotor behavior in the immediate future. A major technical advance in the last decade has been that of obtaining a great deal more valuable information concerning the brain function of each animal used experimentally (Darian-Smith et al. 1990a; Galea and Darian-Smith 1994), resulting in a marked reduction in the number of animals used. We see this increased efficacy in the experimental use of macaques as a continuing trend.

6 References

Acuna C, Cudeiro J, Gonzalez F, Alonso JM, Perez R (1990) Lateral-posterior and pulvinar reaching cells – comparison with parietal area 5a: a study in behaving Macaca nemestrina monkeys. Exp Brain Res 82:158–166

Akbarian S, Grusser OJ, Guldin WO (1992) Thalamic connections of the vestibular cortical fields in the squirrel monkey (Saimiri sciureus). J Comp Neurol 326:423–441

Akbarian S, Grusser OJ, Guldin WO (1994) Corticofugal connections between the cerebral cortex and brainstem vestibular nuclei in the macaque monkey. J Comp Neurol 339:421–437

Allendoerfer KL, Shatz CJ (1994) The subplate, a transient neocortical structure its role in the development of connections between thalamus and cortex. Annu Rev Neurosci 17:185–218

Andersen RA (1989) Visual and eye movement functions of the posterior parietal cortex. Annu Rev Neurosci 12:377–403

Andersen RA (1995) Encoding of intention and spatial location in the posterior parietal cortex. Cerebral Cortex 5:457–469

Andersen RA, Brotchie PR (1992) Spatial maps versus distributed representations. Behav Brain Sci 15:707–709

Andersen RA, Asanuma C, Essick G, Siegel RM (1990a) Corticocortical connections of anatomically and physiologically defined subdivisions within the inferior parietal lobule. J Comp Neurol 296:65–113

Andersen RA, Snowden RJ, Treue S, Graziano M (1990b) Hierarchical processing of motion in the visual cortex of monkey. Cold Spring Harbor Symp 55:741–748

Andersen RA, Brotchie PR, Mazzoni P (1992) Evidence for the lateral intraparietal area as the parietal eye field. Curr Opin Neurobiol 2:840–846

Andersen RA, Snyder LH, Li C, Stricanne B (1993a) Coordinate transformations in the representation of spatial information. Curr Opin Neurobiol 3:171–176

Andersen RA, Treue S, Graziano M, Snowden RJ, Qian N (1993b) From direction of motion to patterns of motion: hierarchies of motion analysis in the visual cortex. In: Ono T, Squire LR, Raichle ME, Perrett DI, Fukada M (eds) Brain mechanisms of perception and memory: from neuron to behavior. Oxford University Press, Oxford, pp 183–199

Anderson ME, Turner RS (1991) Activity of neurons in cerebellar receiving and pallidal receiving areas of the thalamus of the behaving monkey. J Neurophysiol 66:879–893

Apkarian AV, Hodge CJ (1989a) The primate spinothalamic pathways. I. A quantitative study of the cells of origin of the spinothalamic pathway. J Comp Neurol 288:447–473

Apkarian AV, Hodge CJ (1989b) The primate spinothalamic pathways. II. The cells of origin of the dorsolateral and ventral spinothalamic pathways. J Comp Neurol 288:474–492

Apkarian AV, Hodge CJ (1989c) The primate spinothalamic pathways. III. Thalamic terminations of the dorsolateral and ventral spinothalamic pathways. J Comp Neurol 288:493–511

Arbib MA, Hoff B (1994) Trends in neural modeling for reach to grasp. Adv Psychol 105: 311–344

Arbib MA, Iberall T, Lyons D (1985) Coordinated control programs for movements of the hand. In: Goodwin AW, Darian-Smith I (eds) Hand function and the neocortex. Springer, Berlin Heidelberg New York, pp 111–129 (Experimental brain research, Suppl 10)

Armand J, Edgley SA, Lemon RN, Olivier E (1994) Protracted postnatal development of corticospinal projections from the primary motor cortex to hand motoneurones in the macaque monkey. Exp Brain Res 101:178–182

Asanuma C (1993) Specific distribution of some "nonspecific" afferents upon individual thalamic reticular nucleus neurson. In: Miniacchi D, Molinari M, Macchi G, Jones EG (eds) Thalamic networks for relay and modulation. Pergamon, Oxford, pp 323–336

Asanuma C, Thach WT, Jones EG (1983a) Cytoarchitectonic delineation of the ventral lateral thalamic region in the monkey. Brain Res Rev 5:219–235

Asanuma C, Thach WT, Jones EG (1983b) Distribution of cerebellar terninations and their relation to other afferent terminations in the ventral lateral thalamic region of the monkey. Brain Res Rev 5:237–265

Asanuma C, Thach WT, Jones EG (1983c) Anatomical evidence for segregated focal groupings of efferent cells and their terminal ramifications in the cerebellothalamic pathway of the monkey. Brain Res Rev 5:267–298

Asanuma C, Andersen RA, Cowan WM (1985) The thalamic relations of the caudal inferior parietal lobule and the lateral prefrontal cortex in monkeys: divergent cortical projections from cell clusters in the medial pulvinar nucleus. J Comp Neurol 241:357–381

Asanuma H, Rosen I (1972) Topographical organization of cortical efferent zones projecting to distal forelimb muscles in the monkey. Exp Brain Res 14:243–356

Asanuma H, Zarzecki P, Jankowska E, Hongo T, Marcus S (1979) Projection of individual pyramidal tract neurons to lumbar motor nuclei of the monkey. Exp Brain Res 34:73–89

Baker SN, Olivier E, Lemon RN (1994) Interaction between outputs from different parts of the hand representation of the motor cortex in the conscious monkey. J Physiol (Lond) 476P:P28

Baldissera F, Hultborn H, Illert M (1981) Integration in spinal neuronal systems. In: Brookhart JM, Mountcastle VBS, Brooks VBV (eds) The nervous system, vol 2. Motor control, part 1. American Physiological Society, Bethesda, pp 509–596 (Handbook of physiology, sect 1)

Baleydier C, Maugiere F (1987) Network connectivity between parietal area 7, posterior cingulate cortex and medial pulvinar nucleus: a double fluorescent tracer study in monkey. Exp Brain Res 66:385–393

Baleydier C, Morel A (1992) Segregated thalamocortical pathways to inferior parietal and inferotemporal cortex in macaque monkey. Visual Neurosci 8:391–405

Bastings E, Benoordhout AM, Delwaide PJ (1995) Do ipsilateral corticospinal pathways contribute to motor recovery after stroke. Stroke 26:175

Benevento LA, Fallon JH (1975) The ascending projections of the superior colliculus in the rhesus monkey (Macaca mulatta). J Comp Neurol 160:339–362

Bennett KMB, Lemon RN (1994) The influence of single monkey corticomotoneuronal cells at different levels of activity in target muscles. J Physiol (Lond) 477:291–307

Bentivoglio M, Kuypers GJM, Catsman-Bennevoets CE, Leowe H, Dann O (1980) Two new fluorescent retrograde neuronal tracers which are transported over long distances. Neurosci Lett 18:25–30

Berkley KJ (1980) Spatial relationships between the terminations of somatic sensory and motor pathways in the rostral brainstem of cats and monkeys. I. Ascending somatic sensory inputs to lateral diencephalon. J Comp Neurol 193:283–317

Berkley KJ (1983) Spatial relationships between the terminations of somatic sensory motor pathways in the rostral brainstem of cats and monkeys. II. Cerebellar projections compared with those of the ascending sensory pathways in lateral diencephalon. J Comp Neurol 220:229–251

Bernhard CG, Bohm E, Petersén I (1953) Investigations on the organization of the corticospinal system in monkeys (Macaca mulatta). Acta Physiol Scand 29 [Suppl] 106: 79–105

Betz V (1874) Über die feinere Struktur der Grosshinrinde des Menschen. Zbl Med Wiss 19: 209–213

Biber MP, Kneisley LW, LaVail JH (1978) Cortical neurons projecting to the cervical and lumbar enlargements of the spinal cord in young and adult rhesus monkeys. Exp Neurol 59:492–508

Blatt G, Stoner GR, Andersen RA (1987) The lateral intraparietal area (LIP) in the macaque: associational connections and visual receptive field organization. Soc Neurosci Abstr 13:627

Boivie J (1979) An anatomical reinvestigation of the termination of the spinothalamic tract in the monkey. J Comp Neurol 186:343–370

Bortoff GA, Strick PL (1993) Corticospinal terminations in two new world primates further evidence that corticomotoneuronal connections provide part of the neural substrate for manual dexterity. J Neurosci 13:5105–5118

Bowsher D (1961) The termination of secondary somatosensory neurons within the thalamus of Macaca mulatta: an experimental degeneration study. J Comp Neurol 117:213–227

Broca P (1861) Remarques sur le siège de la faculté de language articulé, suives d'une observation d'amphémie (perte de parole). Bull Soc Anat Paris 36:330–337

Brodmann K (1909) Vergleichende Localisationslehre der Grosshirnrinde. Barth, Leipzig

Brooks VB, Thach WT (1981) Cerebellar control of posture and movement. In: Brookhart JM, Mountcastle VBS, Mountcastle VBV (eds) The nervous system, vol 2. Motor control, part 2. American Physiological Society, Bethesda, pp 877–946 (Handbook of physiology, sect 1)

Brouwer B, Ashby P (1992) Corticospinal projections to lower limb motoneurons in man. Exp Brain Res 89:649–654

Buhl EH, Lubke J (1989) Intracellular Lucifer yellow injection in fixed brain slices combined with retrograde tracing, light and electron microscopy. Neurosci 28:3–16

Burman K (1992) Corticospinal projections to the lumbar spinal cord in mature and infant macaques. Thesis, University of Melbourne

Burton H (1986) Second somatosensory cortex and related areas. In: Jones EG, Peters A (eds) Cerebral cortex, vol 5. Sensory-motor areas and aspects of cortical connectivity. Plenum, New York, pp 31–98

Burton H, Sinclair RJ (1991) Second somatosensory cortical area in macaque monkeys. 2. Neuronal responses to punctate vibrotactile stimulation of glabrous skin on the hand. Brain Res 538:127–135

Burton H, Videen TO, Raichle ME (1993) Tactile vibration-activated foci in insular and parietal-opercular studied with positron emission tomography: mapping the second somatosensory area in humans. Somatosens Motor Res 10:297–308

Buys EJ, Lemon RN, Mantel GWH, Muir RB (1986) Selective facilitation of different hand muscles by single corticospinal neurones in the conscious monkey. J Physiol (Lond) 381:529–549

Caminiti R, Johnson PB (1992) Internal representations of movement in the cerebral cortex as revealed by the analysis of reaching. Cerebr Cortex 2:269–276

Caminiti R, Johnson PB, Galli C, Ferraina S, Burnod Y (1991) Making arm movements within different parts of space: the premotor and motor cortical representation of a co-ordinate system for reaching to visual targets. J Neurosci 11:1182–1197

Campbell AW (1905) Histological studies on the localization of cerebral function. Cambridge University Press, Cambridge

Carpenter MB, Nakano K, Kim R (1976) Nigrothalamic projections in the monkey demonstrated by autoradiographic technics. J Comp Neurol 165:401–415

Cavada C, Goldman-Rakic PS (1989a) Posterior parietal cortex in the rhesus monkey. I. Parcellation of areas based on distinctive limbic and sensory corticocortical connections. J Comp Neurol 287:393–421

Cavada C, Goldman-Rakic PS (1989b) Posterior parietal cortex in rhesus monkey. II. Evidence for segregated corticocortical networks linking sensory and limbic areas with the frontal lobe. J Comp Neurol 287:422–445

Cavada C, Goldman-Rakic PS (1993) Multiple visual areas in the posterior parietal cortex of primates. In: Hicks TP, Molotchnikoff S, Ono I (eds) Progress in brain research. Elsevier, New York, pp 123–137

Cajal S Ramon Y (1909–1911) Histologie du systeme nerveux de l'homme et des vertebres (translated by L. Azoulay). Maloine, Paris

Chang HT (1993) Immunoperoxidase labeling of the anterograde tracer Fluoro-Ruby (tetramethylrhodamine-dextran amine conjugate). Brain Res Bull 30:115–118

Cheema S, Rustioni A, Whitsel BL (1984) Light and electron microscopic evidence for a direct corticospinal projection to superficial laminae of the dorsal horn in cats and monkeys. J Comp Neurol 225:276–290

Cheney PD, Fetz EE (1980) Functional classes of primate corticomotoneuronal cells and their relation to active force. J Neurophysiol 44(4):773–791

Cheney PD, Mewes K, Fetz EE (1988) Encoding of motor parameters by corticomotoneuronal (CM) and rubromotoneuronal (RM) cells producing postspike facilitation of forelimb muscles in the behaving monkey. Behav Brain Res 28:181–191

Cheney PD, Fetz EE, Mewes K (1991a) Neural mechanisms underlying corticospinal and rubrospinal control of limb movements. Prog Brain Res 87:213–252

Cheney PD, Mewes K, Widener G (1991b) Effects on wrist and digit muscle activity from microstimuli applied at the sites of rubromotoneuronal cells in primates. J Neurophysiol 66:1978–1992

Chung JM, Lee KH, Surmeier DJ, Sorkin LS, Kim J, Willis WD (1986) Response characteristics of neurons in the ventral posterior lateral nucleus of the monkey thalamus. J Neurophysiol 56:370–390

Cincotta M, Ragazzoni A, Descisciolo G, Pinto F, Maurri S, Barontini F (1994) Abnormal projection of corticospinal tracts in a patients with congential mirror movements. Brain Res 536:301–304

Clark WEL (1932) The structure and connections of the thalamus. Brain 55:406–470

Clark WEL (1936) The termination of ascending tracts in the thalamus of the macaque monkey. J Anat 71:7–40

Clark WEL, Northfield DWC (1937) The cortical projection of the pulvinar in the macaque monkey brain. Brain 60:126–142

Cohen DAD, Prudhomme MJL, Kalaska JF (1994) Tactile activity in primate primary somatosensory cortex during active arm movements correlation with receptive field properties. J Neurophysiol 71:161–172

Colebatch JG, Deiber MP, Passingham RE, Friston KJ, Frackowiak RSJ (1991) Regional cerebral blood flow during voluntary arm and hand movements in human subjects. J Neurophysiol 65:1392–1401

Condé F (1987) Further studies on the use of the fluorescent tracers Fast Blue and Diamidino Yellow: effective uptake area and cellular storage sites. J Neurosci Methods 21:31–44

Corbetta M, Miezin FM, Shulman GL, Petersen SE (1993) A PET study of visuospatial attention. J Neurosci 13:1202–1226

Costello MB, Fragaszy DM (1988) Prehension in Cebus and Saimiri. I. Grip type and hand preference. Am J Primatol 15:235–245

Coulter JD, Jones EG (1977) Differential distribution of corticospinal projections from individual cytoarchitectonic fields in the monkey. Brain Res 129:335–340

Cowey A, Stoerig P, Bannister M (1994) Retinal ganglion cells labeled from the pulvinar nucleus in macaque monkeys. Neurosci 61:691–705

Crammond DJ, Kalaska JF (1989) Neuronal activity in primate parietal cortex area 5 varies with intended movement direction during an instructed-delay period. Exp Brain Res 76:458–462

Critchley M (1953) The parietal lobes. Arnold, London

Crouch RL (1934) The nuclear configuration of the thalamus of Macacus rhesus. J Comp Neurol 59:451–485

Crouch RL, Thompson JK (1938) The efferent fibers of the thalamus of Macacus rhesus. J Comp Neurol 69:255–271

Cusick CG, Gould HJ (1990) Connections between area 3b of the somatosensory cortex and subdivisions of ventroposterior nuclear complex and the anterior pulvinar nucleus in squirrel monkeys. J Comp Neurol 292:83–102

Cusick CG, Steindler DA, Kaas JH (1985) Corticocortical and collateral thalamocortical connections of postcentral somatosensory cortical areas in squirrel monkeys: a double-labeling study with radiolabeled wheatgerm agglutinin and wheatgerm agglutinin conjugated to horseradish peroxidase. Somatosens Res 3(1):1–31

Cusick CG, Scripter JL, Darensbourg JG, Weber JT (1993) Chemoarchitectonic subdivisions of the visual pulvinar in monkeys and their connectional relations with the middle temporal and rostral dorsolateral visual areas, Mt and Dlr. J Comp Neurol 336:1–30

Darian-Smith C, Darian-Smith I (1993) Thalamic projections to area 3a, area 3b, and area 4 in the sensorimotor cortex of the mature and infant macaque monkey. J Comp Neurol 335:173–199

Darian-Smith C, Darian-Smith I, Cheema SS (1990a) Thalamic projections to sensorimotor cortex in the macaque monkey: use of multiple fluorescent tracers. J Comp Neurol 299:17–46

Darian-Smith C, Darian-Smith I, Cheema SS (1990b) Thalamic projections to sensorimotor cortex in the newborn macaque. J Comp Neurol 299:47–63

Darian-Smith C, Darian-Smith I, Burman K, Ratcliffe N (1993a) Ipsilateral cortical projections to areas 3a, 3b and 4 in the sensorimotor cortex of the macaque monkey. J Comp Neurol 335:200–213

Darian-Smith I (1982) Touch in primates. Annu Rev Psychol 33:157–194

Darian-Smith I (1984) The sense of touch: performance and peripheral neural processes. In: Darian-Smith IV, Brookhart JM, Mountcastle VBS (eds) The nervous system, vol 3. American Physiological Society, Bethesda, pp 739–788 (Handbook of physiology, sect 1)

Darian-Smith I, Darian-Smith C (1991a) Distribution of thalamic input to the sensorimotor cortex of the macaque monkey. In: Franzén O, Westman J (eds) Information processing in the somatosensory system. Macmillan, New York, pp 155–173 (Wenner-Gren international symposium series, vol 57)

Darian-Smith I, Sugitani M, Heywood J, Karita K, Goodwin AW (1982) Touching textured surfaces: cells in somatosensory cortex respond both to finger movement and to surface features. Science 281:906:909

Darian-Smith I, Sugitani M, Heywood J, Karita K, Goodwin AW (1984) The tangible features of textured surfaces: their representation in the monkey's somatosensory cortex. In: Edelman GM, Gall WE, Cowan WM (eds) Dynamic aspects of neocortical function. Wiley, New York, pp 475–500

Darian-Smith I, Goodwin A, Sugitani M, Heywood J (1985) Scanning a textured surface with the fingers: events in the sensorimotor cortex. In: Goodwin AW, Darian-Smith I (eds) Hand function and the neocortex. Springer, Berlin Heidelberg New York, pp 17–43

Darian-Smith I, Cheema SS, Darian-Smith C, Galea M, Ratcliffe N (1990) Reaching and handling: how the brain interacts with the hand. In: Aitken LM, Rowe M (eds) Information processing in the mammalian auditory and tactile systems. Liss, New York, pp 191–203

Darian-Smith I, Darian-Smith C, Galea M, Pepperell R (1991b) Touching, handling and using objects: the role of the sensorimotor cortex. In: Huimphrey DR, Freund H-J (eds) Motor control: concepts and issues. Dahlem workshop report. Wiley, Chichester, pp 181–197

Darian-Smith I, Darian-Smith C, Galea MP, Burman K, Tippayatorn N (1993b) Neuron populations in sensorimotor thalamic space: connections, parcellation, and relation to corticospinal projections in the macaque monkey. In: Minciacchi D, Molinari M, Macchi G, Jones EG (eds) Thalamic networks for relay and modulation. Pergamon, Oxford, pp 123–134

Darling WG, Cole KJ, Miller GF (1994) Coordination of index finger movements. J Biomech 27:479–491

Decety J, Kawashima R, Gulyas B, Roland PE (1992) Preparation for reaching a pet study of the participating structures in the human brain. NeuroReport 3:761–764

Deiber MP, Passingham RE, Colebatch JG, Friston KJ, Nixon PD, Frackowiak RSJ (1991) Cortical areas and the selection of movement a study with positron emission tomography. Exp Brain Res 84:393–402

Dodd J, Jessell T (1988) Axon guidance and the patterning of neuronal projections in vertebrates. Science 242:692–699

Donnan GA, Bladin PF, Berkovic SF, Longley WA, Saling MM (1991) The stroke syndrome of striatocapsular infarction. Brain 114:51–70

Dum RP, Strick PL (1989) Corticospinal projections from the motor areas in the frontal lobe. In: Ito M (ed) Neural programming. Japanese Science Society, Tokyo, pp 49–63 (Taniguchi symposia on brain sciences, no 12)

Dum RP, Strick PL (1991) The origin of corticospinal projections from the premotor areas in the frontal lobe. J Neurosci 11:667–689

Dum RP, Strick PL (1992) Medial-wall motor areas and skeletomotor control. Curr Opin Neurobiol 2:836–839

Eccles JC (1964) The physiology of synapses. Springer, Berlin Heidelberg New York

Evarts EV (1966) Pyramidal tract activity associated with a conditioned hand movement in the monkey. J Neurophysiol 29:1011–1027

Evarts EV, Shinoda Y, Wise SP (1984) Neurophysiological approaches to higher brain functions. Neurosciences Institute, New York

Faugier-Grimaud S, Frenois C, Peronnet F (1985) Effects of posterior parietal lesions on visually guided movements in monkeys. Exp Brain Res 59:125–138

Felix D, Wiesendanger M (1971) Pyramidal and non-pyramidal motor cortical effects on distal forelimb muscles of monkeys. Exp Brain Res 12:81–91

Felleman DJ, Van Essen DC (1991) Distributed hierarchical processing in the primate cerebral cortex. Cerebr Cortex 1:1–47

Fenelon G, Francois C, Percheron G, Yelnik J (1990) Topographic distribution of pallidal neurons projecting to the thalamus in macaques. Brain Res 520:27–35

Fenelon G, Yelnik J, Francois C, Percheron G (1994) Central complex of the primate thalamus a quantitative analysis of neuronal morphology. J Comp Neurol 342:463–479

Ferrier D (1876) The functions of the brain. Smith Elder, London

Ferrier D (1890) The Croonian lectures on cerebral localization. Smith Elder, London

Fetz EE (1992) Are movement parameters recognizably coded in the activity of single neurons. Behav Brain Sci 15:679–690

Fetz EE, Cheney PD (1980) Postspike facilitation of forelimb muscle activity by primate corticomotoneuronal cells. J Neurophysiol 44:751–771

Fetz EE, Cheney PD, German DC (1976) Corticomotoneuronal connections of precentral cells detected by postspike averages of EMG activity in behaving monkeys. Brain Res 114: 505–510

Fetz EE, Cheney PD, Mewes K, Palmer S (1989) Control of forelimb muscle activity by populations of corticomotoneuronal and rubromotoneuronal cells. In: Allum JHJ, Hulliger M (eds) Afferent control of posture and locomotion. Elsevier, Amsterdam, pp 437–449 (Progress in brain research, vol 80)

Flament D, Goldsmith P, Lemon RN (1992a) The development of corticospinal projections to tail and hindlimb motoneurons studied in infant macaques using magnetic brain stimulation. Exp Brain Res 90:225–228

Flament D, Hall EJ, Lemon RN (1992b) The development of corticomotoneuronal projections investigated using magnetic brain stimulation in the infant macaque. J Physiol (Lond) 447:755–768

Foerster O, Gagel O (1932) Die Vorderseitenstrangdurchschneidung beim Menschen: eine klinische patholophysiologisch-anatomische Studie. Z Gesamte Neurol Psychiatr 138:1–92 (abstr)

Fredrickson JM, Rubin AM (1986) Vestibular cortex. In: Jones EG, Peters A (eds) Cerebral cortex, vol 5. Sensory-motor areas and aspects of cortical connectivity. Plenum, New York, pp 99–111

Freire M (1990) HRP-filling of neurons and axonal arbors in fixed brain slabs. J Neurosci Methods 35:267–275

Friedemann M (1911) Die Cytoarchitektonik des Zwischenhirns der Cercopitheken mit besonderer Berücksichtigung des Thalamus opticus. J Psychol Neurol 18:309–378

Friedman DP, Jones EG (1981) Thalamic input to areas 3a and 2 in monkeys. J Neurophysiol 25:59–85

Friston KJ, Frith CD, Liddle PF, Frackowiak RSJ (1993) Functional connectivity the principal component analysis of large (PET) data sets. J Cereb Blood Flow Metab 13:5–14

Fromm C, Wise SP, Evarts EV (1984) Sensory response properties of pyramidal tract neurons in the precentral motor cortex and postcentral gyrus of the rhesus monkey. Exp Brain Res 54:177–185

Fulton JF (1949) Physiology of the nervous system. Oxford University Press, New York

Fulton JF, Sheehan D (1935) The uncrossed lateral pyramidal tract in higher primates. J Anat 69:181–187

Fyffe REW (1990) Evidence for separate morphological classes of Renshaw cells in the cat's spinal cord. Brain Res 536:301–304

Galea MP, Darian-Smith I (1994) Multiple corticospinal neuron populations in the macaque monkey are specified by their unique cortical origins, spinal terminations and connections. Cerebr Cortex 4:166–194

Galea MP, Darian-Smith I (1995) Postnatal maturation of corticospinal projections in the macaque. Cerebr Cortex 5

Gall FJ, Spurzheim JC (1809) Recherches sur le système nerveux en générale et celui du cerveux en particulier. Schoell and Nicholle, Paris

Garraghty PE, Sur M (1990) Morphology of single intracellularly stained axons terminating in area 3b of macaque monkeys. J Comp Neurol 294:583–593

Garraghty PE, Pons TP, Sur M, Kaas JH (1989) The arbors of axons terminating in middle cortical layers of somatosensory area 3b in owl monkeys. Somatosens Mot Res 6:401–411

Gatter KC, Powell TPS (1978) The intrinsic connections of the cortex of area 4 of the monkey. Brain 101:513–541

Georgopoulos AP (1986) On reaching. Annu Rev Neurosci 9:147–170

Georgopoulos AP (1991) Higher order motor control. Annu Rev Neurosci 14:361–378

Georgopoulos AP, Caminiti R, Kalaska JF, Massey JT (1983) Spatial coding of movement: a hypothesis concerning the coding of movement direction by motor cortical populations. Exp Brain Res [Suppl] 7:327–336

Georgopoulos AP, Kalaska JF, Crutcher MD, Caminiti R, Massey JT (1984) The representation of movement direction in the motor cortex: single cell and population studies. In: Edelman GM, Gall WE, Cowan WM (eds) Dynamic aspects of neocortical function. Wiley, New York, pp 501–524

Georgopoulos AP, Taira M, Lukashin A (1993) Cognitive neurophysiology of the motor cortex. Science 260:47–52

Gerfen CR (1992) The neostriatal mosaic: multiple levels of compartmental organization in the basal ganglia. Annu Rev Neurosci 15:285–320

Ghosh S, Porter R (1988a) Morphology of pyramidal neurones in monkey motor cortex and the synaptic actions of their intracortical axon collaterals. J Physiol (Lond) 400:593–615

Ghosh S, Porter R (1988b) Corticocortical synaptic influences on morphologically identified pyramidal neurones in the motor cortex of the monkey. J Physiol (Lond) 400:617–629

Ghosh S, Brinkman C, Porter R (1987) A quantitative study of the distribution of neurons projecting to the precentral motor cortex in the monkey (M. fascicularis). J Comp Neurol 259:424–444

Ghosh S, Fyffe REW, Porter R (1988) Morphology of neurons in area 4g of the cat's cortex studied with intracellular injection of HRP. J Comp Neurol 269:290–312

Ghosh A, Antonini A, McConnell SK, Shatz CJ (1990) Requirement for subplate neurons in the formation of thalamocortical connections. Nature 347:179–181

Giguere M, Goldman-Rakic PS (1988) Mediodorsal nucleus: areal, laminar, and tangential distribution of afferents and efferents in the frontal lobe of rhesus monkeys. J Comp Neurol 277:195–213

Goldman-Rakic PS (1987a) Motor control function of the prefrontal cortex. In: Bock G, O'Connor M, Marsh J (eds) Motor areas of the cerebral cortex. Wiley, Chichester, pp 187–197 (Ciba Foundation symposium 132)

Goldman-Rakic PS (1987b) Development of cortical circuitry and cognitive function. Child Dev 58:601–622

Goldman-Rakic PS (1988a) Topography of cognition: parallel distributed networks in primate association cortex. Annu Rev Neurosci 11:137–156

Goldman-Rakic PS (1988b) Changing concepts of cortical connectivity: parallel distributed cortical networks. In: Rakic P, Singer W (eds) Neurobiology of neocortex. Dahlem Workshop Report. Wiley, Chichester, pp 177–202

Goldman-Rakic PS (1993) Neocortical memory cells and circuits. Int Congr Ser 1036:271–280

Goldman-Rakic PS (1994) Specification of higher cortical functions. In: Broman SH, Grafman J (eds) Atypical cognitive deficits I. Lawrence Erlbaum, Hillsdale, pp 3–17

Goldman-Rakic PS, Bates JF, Chafee MV (1992) The prefrontal cortex and internally generated motor acts. Curr Opin Neurobiol 2:830–835
Gowers WR (1886) On the antero-lateral ascending tract of the spinal cord. Lancet 1:1153–1154
Grafton ST, Mazziotta JC, Woods RP, Phelps ME (1992) Human functional anatomy of visually guided finger movements. Brain 115:565–587
Grafton ST, Woods RP, Mazziotta JC (1993) Within arm somatotopy in human motor areas determined by positron emission tomography imaging of cerebral blood flow. Exp Brain Res 95:172–176
Grantyn A, Olivier E, Kitama T (1993) Tracing premotor brain stem networks of orienting movements. Curr Opin Neurobiol 3:973–981
Guillery RW (1966) A study of Golgi preparations from the dorsal lateral geniculate nucleus of the adult cat. J Comp Neurol 128:21–50
Hardy SGP, Lynch JC (1992) The spatial distribution of pulvinar neurons that project to two subregions of the inferior parietal lobule in the macaque. Cerebr Cortex 2:217–230
Harting JK, Noback CR (1970) Corticospinal projections from the pre- and post-central gyri in the squirrel monkey (Saimiri sciureus). Brain Res 24:322–328
Havton LA, Ohara PT (1993a) Quantitative analyses of intracellularly characterized and labeled thalamocortical projection neurons in the ventrobasal complex of primates. J Comp Neurol 336:135–150
Havton LA, Ohara PT (1993b) Morphological studies of a local circuit neuron (LCN) in the primate somatosensory thalamus. Soc Neurosci Abstr 19(1):515
Havton LA, Ohara PT (1994) Dendritic orientation of thalamocortical projection neurons in the ventrobasal complex of macaques. Brain Res 638:126–132
Hayes NL, Rustioni A (1980) Spinothalamic and spinomedullary neurons in macaques: a single and double retrograde tracer study. Neuroscience 5:861–874
Head H (1918) Sensation and the cerebral cortex. Brain 41:57–253
Head H, Holmes G (1911) Sensory disturbances from cerebral lesions. Brain 34:102–254
Heffner CD, Lumsden AGS, O'Leary DDM (1990) Target control of collateral extension and directional axon growth in the mammalian brain. Science 247:217–220
Hendry SHC, Jones EG (1983) Thalamic inputs to identified commissural neurons in the monkey somatic sensory cortex. J Neurocytol 12:299–316
Hendry SHC, Jones EG, Emson PC, Lawson DEM, Heizmann DW, Streit P (1989) Two classes of cortical GABA neurons defined by differential calcium binding protein immunoreactivities. Exp Brain Res 76:467–472
Hinton GE, McClelland JL, Rumelhart DE (1986) Distributed representations. In: Rumelhart DE, McClelland JL (eds) Parallel distributed processing, vol 1. MIT Press, Cambridge, MA, pp 77–109
Hirai T, Jones EG (1989) A new parcellation of the human thalamus on the basis of histochemical staining. Brain Res Rev14:1–34
Hirai T, Schwark HD, Yen C-T, Honda CN, Jones EG (1988) Morphological features of physiologically characterized medial lemniscal axons in cat ventral posterior thalamic nucleus. J Neurophysiol 60:1439–1459
Holmes G (1926) Disorders of sensation produced by cortical lesions. Brain 50:413–427
Holmes G, May WP (1909) On the exact origin of the pyramidal tracts in man and other mammals. Brain 32(1):1–43
Holsapple JW, Preston JB, Strick PL (1991) The origin of thalamic inputs to the hand representation in the primary motor cortex. J Neurosci 11:2644–2654
Holstege G, Blok BF, Ralston DD (1988) Anatomical evidence for red nucleus projections to motoneuronal cell groups in the spinal cord of the monkey. Neurosci Lett 95:97–101
Hore J, Preston JB, Durkovic RG, Cheney PD (1976) Responses of cortical neurons (areas 3a and 4) to ramp stretch of hindlimb muscles in the baboon. J Neurophysiol 39: 484–500
Houk JC, Keifer J, Barto AG (1993) Distributed motor commands in the limb premotor network. TINS 16:27–33
Houser CR, Hendry SHC, Jones EG, Vaughn JE (1983) Morphological diversity of immunocytochemically identified GABA neurons in the monkey sensory-motor cortex. J Neurocytol 12:617–638

Hultborn H, Illert M (1991) How is motor behaviour reflected in the organization of spinal systems? In: Humphrey DR, Freund H (eds) Motor control: concepts and issues. Dahlem workshop report. Wiley, Chichester, pp 49–73

Humphrey DR, Tanji J (1991) What features of voluntary movement are encoded in the neuronal discharge of different cortical motor areas? In: Humphrey DR, Freund H (eds) Motor control: concepts and issues. Dahlem workshop report. Wiley, Chichester, pp 413–443

Humphrey DR, Gold R, Reed DJ (1984) Sizes, laminar and topographic origins of cortical projections to the major divisions of the red nucleus in the monkey. J Comp Neurol 225:75–94

Hutchins KD, Martino AM, Strick PL (1988) Corticospinal projections from the medial wall of the hemisphere. Exp Brain Res 71:667–672

Hyvarinen J (1982) The parietal cortex of monkey and man. Springer, Berlin Heidelberg New York

Iberall T, Bingham G, Arbib MA (1986) Opposition space as a structuring concept for the analysis of skilled movements. In: Heuer H, Fromm C (eds) Generation and modulation of action patterns. Springer, Berlin Heidelberg New York, pp 158–173

Ilinsky IA, Kultas-Ilinsky K (1987) Sagittal cytoarchitectonic maps of the Macaca mulatta thalamus with a revised nomenclature of the motor-related nuclei validated by observations on their connectivity. J Comp Neurol 262:331–364

Ilinsky IA, Kultas-Ilinsky K (1990) Fine structure of the magnocellular subdivision of the ventral anterior thalamic nucleus (VAmc) of Macaca mulatta. I. Cell types and synaptology. J Comp Neurol 294:455–478

Ilinsky IA, Jouandet ML, Goldman-Rakic PS (1985) Organization of the nigrothalamic cortical system in the rhesus monkey. J Comp Neurol 236:315–330

Ilinsky IA, Toga AW, Kultas-Ilinsky K (1993a) Anatomical organization of internal neural circuits in the the motor thalamus. In: Minciacchi D, Molinari M, Macchi G, Jones EG (eds) Thalamic networks for relay and modulation. Pergamon, Oxford, pp 155–164

Ilinsky IA, Tourtellotte WG, Kultas-Ilinsky K (1993b) Anatomical distinctions between the two basal ganglia afferent territories in the primate motor thalamus. Stereotact Funct Neurosurg 60:62–69

Itaya SK, Van Hoesen GW (1983) Retinal projections to the inferior and medial pulvinar nuclei in the old-world monkey. Brain Res 269:223–230

Ito M (1984) The cerebellum and motor control. Raven, New York

Iwamura Y (1993) Dynamic and hierarchical processing in the monkey somatosensory cortex. Biomed Res 14:107–111

Iwamura Y, Tanaka M (1978) Postcentral neurons in the hand region of area 2: their possible role in the form discrimination of tactile objects. Brain Res 150:662–666

Jackson JH (1875) On the anatomical and physiological localization of movements in the brain. In: Taylor J (ed) Selected writings of John Hughlings Jackson, vol 1. On epilepsy and epileptiform convulsion. Hodder and Stoughton, London

Jackson JH (1987) Relations of different divisions of the central nervous system to one another and to parts of the body. In: Taylor J (ed) Selected writings of John Hughlings Jackson, vol 2. Evolutions and dissolution of the nervous system, speech. Various addresses and lectures. Hodder and Stoughton, London

Jankowska E, Edgley S (1993) Interactions between pathways controlling posture and gait at the level of spinal interneurons in the cat. Annu Rev Neurosci 15:403–442

Jeannerod M (1985) The posterior parietal area as a spatial generator. In: Ingle DJ, Jeannerod M, Lee DN (eds) Brain mechanisms and spatial vision. Nijhoff, Dordrecht, pp 279–298

Jeannerod M (1988) The neural and behavioural organization of goal-directed movements. Clarendon, Oxford

Jeannerod M (1994a) The representing brain neural correlates of motor intention and imagery. Behav Brain Sci 17:187–202

Jeannerod M (1994b) The hand and the object the role of posterior parietal cortex in forming motor representations. Can J Physiol Pharmacol 72:535–541

Jeannerod M, Decety J, Michel F (1994) Impairment of grasping movements following a bilateral posterior parietal lesion. Neuropsychologia 32:369–380

Jinnai K, Nambu A, Tanibuchi I, Yoshida S (1993) Cerebellothalamic and pallido thalamic pathways to area 6 and area 4 in the monkey. Stereotact Funct Neurosurg 60:72–79

Jones EG (1975) Some aspects of the organization of the thalamic reticular complex. J Comp Neurol 162:285–308

Jones EG (1983a) Thalamic basis for column-like input to monkey somatic sensory and motor cortex. In: Macchi G, Rustioni A, Spreatico R (eds) Somatosensory integration in the thalamus. Elsevier Science, Amsterdam, pp 309–336

Jones EG (1983b) Lack of collateral thalamocortical projections to fields of the first somatic sensory cortex in monkeys. Exp Brain Res 52:375–384

Jones EG (1983c) Distribution patterns of individual medial lemniscal axons in the ventrobasal complex of the monkey thalamus. J Comp Neurol 215:1–16

Jones EG (1985) The thalamus. Plenum, New York

Jones EG (1987a) New insights into old concepts of thalamic function. In: Besson JM, Guilbaud G, Peschanski M (eds) Thalamus and pain. Elsevier, Amsterdam, pp 1–19

Jones EG (1987b) Immunocytochemical studies on thalamic afferent transmitters. In: Besson JM, Guilbaud G, Peschanski M (eds) Thalamus and pain. Elsevier, Amsterdam, pp 83–110

Jones EG (1987c) Ascending inputs to, and internal organization of, cortical motor areas. Ciba Found Symp 132:21–35

Jones EG (1989) Defining the thalamic intralaminar nuclei in primates. In: Gainotti G, Bentivoglio M, Bergonzi P, Ferro FM (eds) Neurologia e scienze di base: scritti in onore di Giorgio Macchi. Universita Cattolica del Sacro Cuore, Milan, pp 161–193

Jones EG, Friedman DP (1982) Projection pattern of functional components of thalamic ventrobasal complex on monkey somatosensory cortex. J Neurophysiol 48:521–544

Jones EG, Powell TPS (1970a) Connexions of the somatic sensory cortex of the rhesus monkey. III. Thalamic connexions. Brain 93:37–56

Jones EG, Powell TPS (1970b) An Anatomical study of converging sensory pathways within the cerebral cortex of the monkey. Brain 93:793–820

Jones EG, Coulter JD, Hendry SHC (1978) Intracortical connectivity of architectonic fields in the somatic sensory, motor and parietal cortex of monkeys. J Comp Neurol 181:291–348

Jones EG, Friedman DP, Hendry SHC (1982) Thalamic basis of place- and modality-specific columns in monkey somatosensory cortex: a correlative anatomical and physiological study. J Neurophysiol 48:545–568

Kaas JH, Nelson RJ, Sur M, Lin C, Merzenich MM (1979) Multiple representations of the body within the primary somatosensory cortex of primates. Science 204:521–523

Kaas JH, Nelson RJ, Sur M, Merzenich MM (1981) Organization of somatosensory cortex in primates. In: Schmitt FO, Worden FG, Adelman G, Dennis SG (eds) The organization of the cerebral cortex. MIT Press, Cambridge, pp 237–261

Kaas JH, Nelson RJ, Sur M, Dykes RW, Merzenich MM (1984) The somatotopic organization of the ventroposterior thalamus of the squirrel monkey, Saimiri sciureus. J Comp Neurol 226:111–140

Kalaska JF, Crammond DJ (1992) Cerebral cortical mechanisms of reaching movements. Science 255:1517–1523

Kalaska JF, Hyde ML (1985) Area 4 and area 5: differences between the load direction-dependent discharge variability of cells during active postural fixation. Exp Brain Res 59:197–202

Kalaska JF, Caminiti R, Georgopoulos AP (1983) Cortical mechanisms related to the direction of two-dimensional arm movements: relations in parietal area 5 and comparison with motor cortex. Exp Brain Res 51:247–260

Kalaska JF, Cohen DAD, Hyde ML, Prud'homme M (1989) A comparison of movement-related versus load direction-related activity in primate motor cortex, using a two-dimensional reaching task. J Neurosci 9:2080–2102

Kalaska JF, Cohen DAD, Prud'homme M, Hyde ML (1990) Parietal area 5 neuronal activity encodes movement kinematics not movement dynamics. Exp Brain Res 80:351–364

Kalil K (1981) Projections of the cerebellar and dorsal column nuclei upon the thalamus of the rhesus monkey. J Comp Neurol 195:25–50

Kalil K (1984) Development and regrowth of the rodent pyramidal tract. TINS 394–398
Katz LC, Iarovici DM (1990) Green fluorescent latex microspheres – a new retrograde tracer. Neuroscience 34:511–520
Katz LC, Burkhalter A, Dreyer WJ (1984) Fluorescent latex microspheres as a retrograde marker for in vivo and in vitro studies of visual cortex. Nature 310:498–500
Kavanagh M (1983) A complete guide to monkeys, apes and other primates. Cape, London
Kawashima R, Yamada K, Kinomura S, Yamaguchi T, Matsui H, Yoshioka S, Fukuda H (1993) Regional cerebral blood flow changes of cortical motor areas and prefrontal areas in humans related to ipsilateral and contralateral hand movement. Brain Res 623:33–40
Keifer J, Houk JC (1994) Motor function of the cerebellorubrospinal system. Physiol Rev 74:509–542
Keizer K, Kuypers HGJM (1989) Distribution of corticospinal neurons with collaterals to the lower brain stem reticular formation in monkey (Macaca fascicularis). Exp Brain Res 74:311–318
Keizer K, Kuypers HGJM, Huisman AM, Dann O (1983) Diamidino yellow dihydrochloride (DY-2HCL); a new fluorescent retrograde neuronal tracer, which migrates only very slowly out of the cell. Exp Brain Res 51:179–191
Kennard MA (1938) Reorganization of motor function in the cerebral cortex of monkeys deprived of motor and premotor areas in infancy. J Neurophysiol 1:477–496
Kennard MA (1940) Relation of age to motor impairment in man and subhuman primates. Arch Neurol Psychiatr 44:377–397
Kennedy H, Dehay C (1993) Cortical specification of mice and men. Cerebr Cortex 3:171–186
Kennedy PR (1990) Corticospinal, rubrospinal and rubro-olivary projections: a unifying hypothesis. TINS 13:474–479
Kennedy PR, Gibson AR, Houk JC (1986) Functional and anatomic differentiation between parvicellular and magnocellular regions of red nucleus in the monkey. Brain Res 364: 124–136
Kievit J, Kuypers HGJM (1977) Organization of thalamo-cortical connexions to the frontal lobe in the rhesus monkey. Exp Brain Res 29:299–322
Kim GJ, Shatz CJ, McConnell SK (1991) Morphology of pioneer and follower growth cones in the developing cerebral cortex. J Neurobiol 22:629–642
Knudsen EI, Brainard MS (1995) Creating a unified representation of visual and auditory space in the brain. Annu Rev Neurosci 18:19–44
Krieg WJS (1948) A reconstruction of the diencephalic nuclei of Macacus rhesus. J Comp Neurol 88:1–52
Kultas-Ilinsky K, Ilinsky IA (1991) Fine structure of the ventral lateral nucleus (VL) of the Macaca mulatta thalamus: cell types and synaptology. J Comp Neurol 314:319–349
Kuypers HGJM (1962) Corticospinal connections: postnatal development in the rhesus monkey. Science 138:678–680
Kuypers HGJM (1981) Anatomy of the descending pathways. In: Brooks VBV, Brookhart JM, Mountcastle VBS (eds) The nervous system, vol 2. American Physiological Society, Bethesda, pp 597–666 (Handbook of physiology, sect 1)
Kuypers HGJM, Brinkman J (1970) Precentral projections to different parts of the spinal intermediate zone in the rhesus monkey. Brain Res 24:29–48
Kuypers HGJM, Lawrence DG (1967) Cortical projections to the red nucleus and the brain stem in the rhesus monkey. Brain Res 4:151–188
Kuypers HGJM, Bentivoglio M, Catsman-Berrevoets CE, Bharos AT (1980) Double retrograde labeling through divergent axon collaterals, using two fluorescent tracers with the same excitation wavelength which label different features of the cell. Exp Brain Res 40:383–392
Lachica EA, Beck PD, Casagrande VA (1992) Parallel pathways in macaque monkey striate cortex: anatomically defined columns in layer III. Proc Natl Acad Sci USA 89:3566–3570
LaMotte RH, Mountcastle VB (1979) Disorders in somaesthesis following lesions of parietal lobe. J Physiol (Paris) 42:400–419
Larkman AU (1991a) Dendritic morphology of pyramidal neurones of the visual cortex of the rat: I. Branching patterns. J Comp Neurol 306:307–319

Larkman AU (1991b) Dendritic morphology of pyramidal neurones of the visual cortex of the rat. II. Parameter correlations. J Comp Neurol 306:320–331

Larkman AU (1991c) Dendritic morphology of pyramidal neurons of the visual cortex of the rat. III. Spine distributions. J Comp Neurol 306:332–343

Lawrence DG (1994) Central neural mechanisms of prehension. Can J Physiol Pharamacol 72:580–582

Lawrence DG, Kuypers HGJM (1968a) The functional organization of the motor system in the monkey. I. The effects of bilateral pyramidal lesions. Brain 91:1–14

Lawrence DG, Kuypers HGJM (1968b) The functional organization of the motor system in the monkey. II. The effects of lesions of the descending brain-stem pathways. Brain 91: 15–33

Lawrence DG, Porter R, Redman SJ (1985) Corticomotoneuronal synapses in the monkey: light microscopical localization upon motoneurons of intrinsic muscles of the hand. J Comp Neurol 232:499–510

Leichnetz GR (1986) Afferent and efferent connections of the dorsolateral precentral gyrus (area 4, hand/arm region) in the macaque monkey, with comparisons to area 8. J Comp Neurol 254:460–492

Lemon RN (1993) Cortical control of the primate hand. Exp Physiol 78:263–301

Lemon RN, Mantel GWH, Muir RB (1986) Corticospinal facilitation of hand muscles during voluntary movement in the conscious monkey. J Physiol 381:497–527

LeVay S, Ferster D (1977) Relay cell classes in the lateral geniculate nucleus of the cat and the effects of deprivation. J Comp Neurol 172:563–584

Leventhal AG, Rodieck RW, Dreher B (1981) Retinal ganglion cell classes in the Old World monkey: morphology and central projections. Science 213:1139–1142

Leyton ASF, Sherrington CS (1917) Observations on the excitable cortex of the chimpanzee, orangutan and gorilla. Q J Exp Physiol 11:135–222

Liu CN, Chambers WW (1964) An experimental study of the corticospinal system in the monkey (Macaca mulatta). The spinal pathways and preterminal distribution of degenerating fibers following discrete lesions of the pre- and postcentral gyri and bulbar pyramid. J Comp Neurol 123:257–284

Liu X, Warren RA, Jones EG (1995a) Synaptic distribution of afferents from reticular nucleus in ventrobasal nucleus of cat thalamus. J Comp Neurol 352:187–202

Liu X, Honda CN, Jones EG (1995b) Distribution of four types of synapse on physiologically identified relay neurons in the ventral posterior thalamic nucleus of the cat. J Comp Neurol 352:69–91

Livingstone M, Hubel D (1988) Segregation of form, color, movement, and depth: anatomy, physiology, and perception. Science 240:740–749

Loe PR, Whitsel BL, Dreyer DA, Metz CB (1977) Body representation in ventrobasal thalamus of macaque: a single-unit analysis. J Neurophysiol 40:1339–1355

Lubke J (1993) Morphology of neurons in the thalamic reticular nucleus (TRN) of mammals as revealed by intracelluler injections into fixed tissue slices. J Comp Neurol 329:458–471

Lundberg A (1966) Integration in the reflex pathway. In: Granit R (ed) Muscular afferents and motor control. Nobel symposium I. Almqvist and Wiksell, Stockholm, pp 275–305

Lundberg A (1982) Inhibitory control from the brain stem of transmission from primary afferents to motoneurons, primary afferent terminals and ascending pathways

Luppino G, Matelli M, Camarda RM, Gallese V, Rizzolatti G (1991) Multiple representations of body movements in mesial area 6 and the adjacent cingulate cortex: an intracortical microstimulation study in the macaque monkey. J Comp Neurol 311:463–482

Luppino G, Marelli M, Camarda R, Rizzolatti G (1994) Corticospinal projections from mesial frontal and cingulate areas in the monkey. Neuroreport 5:2545–2548

Macchi G (1993) The intralaminar system revisited. In: Minciacchi D, Molinari M, Macchi G, Jones EG (eds) Thalamic networks for relay and modulation. Pergamon, Oxford, pp 175–184

Macchi G, Bentivoglio M (1986) The thalamic intralaminar nuclei and the cerebral cortex. In Jones EG, Peters A (eds) Cerebral cortex, vol 5. Plenum, New York, pp 355–401

MacPherson JM, Marangoz C, Miles TS, Wiesendanger M (1982) Microstimulation of the supplementary motor area (SMA) in the awake monkey. Exp Brain Res 45:410–416

Maier MA, Bennett KMB, Heppreymond MC, Lemon RN (1993) Contribution of the monkey corticomotoneuronal system to the control of force in precision grip. J Neurophysiol 69:772–785

Marr D (1969) A theory of cerebellar cortex. J Physiol (Lond) 202:437–470

Martin JH, Ghez C (1988) Red nucleus and motor cortex: parallel motor systems for the initiation and control of skilled movement. Behav Brain Res 28:217–223

Martin JH, Ghez C (1993) Differential impairments in reaching and grasping produced by local inactivation within the forelimb representation of the motor cortex in the cat. Exp Brain Res 94:429–443

Martin KAC (1992) Parallel pathways converge. Curr Biol 2:555–557

Martino AM, Strick PL (1987) Corticospinal projections originate from the arcuate premotor area. Brain Res 404:307–312

Matelli M, Luppino G (1993) Cortical projections of motor thalamus. In: Minciacchi D, Molinari M, Macchi G, Jones EG (eds) Thalamic networks for relay and modulation. Pergamon, Oxford, pp 165–174

Matelli M, Camarda R, Glickstein M, Rizzolatti G (1984) Interconnections within the postarcuate cortex (area 6) of the macaque monkey. Brain Res 310:388–392

Matelli M, Camarda R, Glickstein M, Rizzolatti G (1986) Afferent and efferent projections of the inferior area 6 in the macaque monkey. J Comp Neurol 251:281–298

Matelli M, Luppino G, Fogassi L, Rizzolatti G (1989) Thalamic input to inferior area 6 and 4 in the macaque monkey. J Comp Neurol 280:468–488

Matelli M, Luppino G, Rizzolatti G (1991) Architecture of superior and mesial area 6 and the adjacent cingulate cortex in the macaque monkey. J Comp Neurol 311:445–462

Matelli M, Rizzolatti G, Bettinardi V, Gilardi MC, Perani D, Rizzo G, Fazio F (1993) Activation of precentral and mesial motor areas during the execution of elementary proximal and distal arm movements: a PET study. Neuroreport 4:1295–1298

Matsuzaka Y, Aizawa H, Tanji J (1992) A motor area rostral to the supplementary motor area (presupplementary motor area) in the monkey. Neuronal activity during a learned motor task. J Neurophysiol 68:653–662

Maunsell JHR, Newxome WT (1987) Visual processin gin monkey extrastriate cortex. Annu Rev Neurosci 10:363–402

McClelland JL (1989) Parallel distributed processing: implications for cognition and development. In: Morris RGM (ed) Parallel distributed processing: implications for psychology and neurobiology. OUP, New York, pp 8–45

McClelland JF, Rumelhart JL, Hinton GE (1986) The appeal of parallel distributed processing. In Rumelhart DE, McClelland JL (eds) Parallel distributed processing, vol 1. MIT Press, Cambridge, MA, pp 3–44

McCloskey DI (1994) Human proprioceptive sensation. J Clin Neurosci 1:173–177

McCormick DA (1992) Neurotransmitter actions in the thalamus and cerebral cortex and their role in neuromodulation of thalamocortical activity. Prog Neurobiol 39:337–388

McCormick DA, Vonkrosigk M (1992) Corticothalamic activation modulates thalamic firing through glutamate metabotropic receptors. Proc Natl Acad Sci USA 89:2774–2778

McCrea DA (1992) Can sense be made of spinal interneuron circuits. Behav Brain Sci 15:633–643

Mehler WR, Feferman ME, Nauta WJ (1960) Ascending axon degeneration following anterolateral cordotomy: experimental study in the monkey. Brain 83:718–750

Merigan WH, Maunsell JHR (1993) How parallel are the primate visual pathways? Annu Rev Neurosci 16:369–402

Mewes K, Cheney PD (1994) Primate rubromotoneuronal cells parametric relations and contribution to wrist movement. J Neurophysiol 72:14–30

Miniacchi D, Bentivoglio M, Molinari M, Kultas-Ilinsky K, Ilinsky IA, Macchi G (1986) Multiple cortical targets of one thalamic nucleus: the projections of the ventral medial nucleus in the cat studied with retrograde tracers. J Comp Neurol 252:106–129

Mitrofanis J, Guillery RW (1993) New views of the thalamic reticular nucleus in the adult and the developing brain. TINS 16:240–245

Mitz AR, Wise SP (1987) The somatotopic organization of the supplementary motor: intracortical microstimulation mapping. J Neurosci 7:1010–1021

Miyata M, Sasaki K (1983) HRP studies on thalamocortical neurons related to the cerebellocerebral projection in the monkey. Brain Res 274:213–224

Miyata M, Sasaki K (1984) Horseradish peroxidase studies on thalamic and striatal connections of the mesial part of area 6 in the monkey. Neurosci Lett 49:127–133

Monmonnier M (1991) How to lie with maps. University of Chicago Press, Chicago

Morecraft RJ, Vanhoesen GW (1992) Cingulate input to the primary and supplementary motor cortices in the rhesus monkey: evidence for somatotopy in areas 24c and 23c. J Comp Neurol 322:471–489

Mountcastle VB (1984) Central nervous mechanisms in mechanoreceptive sensibility. In: Darian-Smith I, Brookhart JM, Mountcastle VB (eds) The nervous system. III. Sensory processes. American Physiological Society, Bethesda, pp 789–878 (Handbook of physiology, sect 1)

Mountcastle VB, Henneman N (1952) The representation of tactile sensibility in the thalamus of the monkey. J Comp Neurol 97:409–440

Mountcastle VB, Poggio GF (1963) The functional properties of ventrobasal thalamic neurons studied in unanesthetized monkeys. J Neurophysiol 26:775–806

Mountcastle VB, Powell TPS (1959) Central nervous mechanisms subserving position sense and kinesthesis. Bull Johns Hopkins Hosp 105:173–200

Mountcastle VB, Lynch JC, Georgopoulos A, Sakata H, Acuna C (1975) Posterior parietal association cortex of the monkey: command functions for operations within extrapersonal space. J Neurophysiol 38:871–908

Mountcastle VB, Lynch JC, Georgopoulos A, Sakata H, Acuna C (1978) Posterior parietal association cortex of the monkey: command functions for operations within extrapersonal space. J Neurophysiol 38:871–908

Muir RB (1985) Small hand muscles in precision grip: a corticospinal prerogative? In: Goodwin AW, Darian-Smith I (eds) Hand function and the neocortex. Springer, Berlin Heidelberg New York, pp 155–174

Muir RB, Lemon RN (1983) Corticospinal neurons with a special role in precision grip. Brain Res 261:312–316

Murray EA, Coulter JD (1981) Organization of corticospinal neurons in the monkey. J Comp Neurol 195:339–365

Mushiake H, Inase M, Tanji J (1991) Neuronal activity in the primate premotor, supplementary, and precentral motor cortex during visually guided and internally determined sequential movements. J Neurophysiol 66:705–718

Nakagawa S, Tanaka S (1984) Retinal projections to the pulvinar nucleus of the macaque monkey: a reinvestigation using autoradiography. Exp Brain Res 57:151–157

Nance DM, Burns J (1990) Fluorescent dextrans as sensitive anterograde neuroanatomical tracers: applications and pitfalls. Brain Res Bull 25:139–145

Napier J (1980) Hands. Allen and Unwin, London

Nathan PW, Smith MC (1982) The rubrospinal and central tegmental tracts in man. Brain 105:223–269

Nathan PW, Smith MC, Deacon P (1990) The corticospinal tracts in man. Course and location of fibres at different segmental levels. Brain 113:303–324

Nelson ME, Bower JM (1990) Brain maps and parallel computers. TINS 13:403–408

Nelson RJ, Sur M, Felleman DJ, Kaas JH (1980) Representation of the body surface in postcentral pareital cortex (SI) of Macaca fascicularis. J Comp Neurol 192:611–643

Nissl F (1913) Die Grosshirnanteile des Kaninchens. Arch Psychiatr Nervenkr 52:867–953

Nudo RJ, Masterton RB (1990) Descending pathways to the spinal cord. III. Sites of origin of the corticospinal tract. J Comp Neurol 296:559–583

Ogren MP (1993) The development of the primate pulvinar

O'Leary D, Stanfield BB (1986) A transient pyramidal tract projection from the visual cortex in the hamster and its removal by selective collateral elimination. Dev Brain Res 27:87–99

O'Leary D, Schlagger BL, Tuttle R (1994) Specification of neocortical areas and thalamocortical connections. Annu Rev Neurosci 17:419–440

Ohara PT, Havton LA (1994) Preserved features of thalamocortical projection neuron dendritic architecture in the somatosensory thalamus of the rat, cat and macaque. Brain Res 648:259–264

Ohara PT, Chazal G, Ralston HJ (1989) Ultrastructural analysis of GABA-immunoreactive elements in the monkey thalamic ventrobasal complex. J Comp Neurol 283:541–558

Ojima H, Honda CN, Jones EG (1991) Patterns of axon collateralization of identified supragranular pyramidal neurons in the cat auditory cortex. Cerebr Cortex 1:45–62

Olszewski J (1952) The thalamus of the Macaca mulatta. Karger, Basel

Padel Y (1993) Magnocellular and parvocellular red nuclei: anatomical and functional aspects of their connections with the cerebellum and other nervous centers. Rev Neurol 149:703–715

Padel Y, Relova JL (1988) A common somaesthetic pathway to red nucleus and motor cortex. Behav Brain Res 28:153–157

Padel Y, Bourbonnais D, Sybirska E (1986) A new pathway from primary afferents to the red nucleus. Neurosci Lett 64:75–80

Pandya DN, Van Hoesen GW, Mesulam MM (1981) Efferent connections of the cingulate gyrus in the rhesus monkey. Exp Brain Res 42:319–330

Payne JN, Peace JM (1989) Fluorometric comparisons of the retrograde axonal transport of true blue and diamidino yellow from the rat caudate-putamen. Exp Brain Res 75:169–182

Pearson RCA, Powell TPS (1985) The projection of the primary somatic sensory cortex upon area 5 in the monkey. Brain Res Rev 9:89–107

Penfield W, Boldrey E (1937) Somatic motor and sensory representation in the cerebral cortex of man as studied by electrical stimulation. Brain 60:389–443

Penfield W, Welch K (1951) The supplementary motor area of the cerebral cortex. Arch Neurol Psychiatr 66:289–317

Percheron G, Francois C, Yelnik J, Fenelon G, Talbi B (1993a) Informational neuromorphology of the cortico cerebello pontine thalamo cortical system in primates (compared with the basal ganglia system). Rev Neurol 149:678–691

Percheron G, Francois C, Yelnik J, Talbi B, Meder JF, Fenelon G (1993b) The pallidal and nigral thalamic teritories and the problem of the anterior part of the lateral region in primates. In: Minciacchi D, Molinari M, Macchi G, Jones EG (eds) Thalamic networks for relay and modulation. Oxford, Pergamon, pp 145–154

Petrides M, Pandya DN (1984) Projections to the frontal cortex from the posterior parietal region in the rhesus monkey. J Comp Neurol 228:105–116

Phillips CG, Porter R (1964) The pyramidal projection to motoneurones of some muscle groups of the baboon's forearm. In: Eccles JC, Schadé JP (eds) Physiology of spinal neurons. Elsevier, Amsterdam, pp 222–243 (Progress in brain research, vol 12)

Phillips CG, Porter R (1977) Corticospinal neurones. Their role in movement. Academic, New York

Poggio GF, Mountcastle VB (1960) A study of the functional contributions of the lemniscal and spinothalamic systems to somatic sensibility. Bull Johns Hopkins Hosp 266–316

Poggio G, Mountcastle VB (1963) The functional properties of the ventrobasal thalamic neurons studied in unanesthetized monkeys. J Neurophysiol 26:775–806

Polyak S (1932) The main afferent fiber systems of the cerebral cortex in primates. University of California Press, Berkeley

Pons TP, Kaas JH (1985) Connections of area 2 of somatosensory cortex with the anterior pulvinar and subdivisions of the ventroposterior complex in macaque monkeys. J Comp Neurol 240:16–36

Pons TP, Garraghty PE, Mishkin M (1992) Serial and parallel processing of tactual information in somatosensory cortex of rhesus monkeys. J Neurophysiol 68:518–527

Porrino LJ, Goldman-Rakic PS (1982) Brainstem innervation of prefrontal and anterior cingulate cortex in the rhesus monkey revealed by retrograde transport of HRP. J Comp Neurol 205:63–76

Porter R (1972) Relationship of the discharges of cortical neurons to movement in free-to-move monkeys. Brain Res 40:39–43

Porter R, Lemon R (1993) Corticospinal function and voluntary movement. Clarendon, Oxford

Porter R, Lewis MM (1975) Relationship of neuronal discharges in the precentral gyrus of monkeys to the performance of arm movements. Br Res 98:21–36

Posner MI (1993) Seeing the mind. Science 262:672–673

Posner MI, Petersen SE (1990) The attention system of the human brain. Annu Rev Neurosci 13:25–42
Powell TPS, Mountcastle VB (1959) The cytoarchitecture of the postcentral gyrus of the monkey Macaca mulatta. Bull Johns Hopkins Hosp 105:108–131
Raichle ME, Fiez JA, Videen TO, Macleod AMK, Pardo JV, Fox PT, Petersen SE (1994) Practice related changes in human brain functional anatomy during nonmotor learning. Cerebr Cortex 4:8–26
Rajakumar N, Elisevich K, Flumerfelt BA (1993) Biotinylated dextran: a versatile anterograde and retrograde neuronal tracer. Brain Res 607:47–53
Rakic P, Bourgeois J, Eckenhoff MF, Zecevic N, Goldman-Rakic PS (1986) Concurrent overproducton of synapses in diverse regions of the primate cerebral cortex. Science 232:232–235
Ralston DD (1994a) Corticorubral synaptic organization in Macaca fascicularis:a study utilizing degeneration, anterograde transport of WGA-HRP, and combined immuno-GABA-gold technique and computer-assisted reconstruction. J Comp Neurol 350:657–673
Ralston DD (1994b) Cerebellar terminations in the red nucleus of macaca fascicularis an electron microscopic study utilizing the anterograde transport of WGA HRP. Somatosens Mot Res 11:101–107
Ralston DD (1994c) Primary motor cortical synaptic afferents in the red nucleus of Macaca fascicularis a quantitative electron microscopic analysis of corticorubral connectivity. FASEB J 8:A889
Ralston DD, Milroy AM (1992) Inhibitory synaptic input to identified rubrospinal neurons in macaca fascicularis an electron microscopic study using a combined immuno-GABA-gold technique and the retrograde transport of WGA-HRP. J Comp Neurol 320:97–109
Ralston DD, Ralston HJ (1985) The terminations of corticospinal tract axons in the macaque monkey. J Comp Neurol 242:325–337
Ralston DD, Milroy AM, Holstege G (1988) Ultrastructural evidence for direct monosynaptic rubrospinal connections to motoneurons in Macaca mulatta. Neurosci Lett 95:102–106
Ralston HJ (1971) Evidence for presynaptic dendrites and a proposal for their method of action. Nature 230:585–587
Ralston HJ (1991) Local circuitry of the somatosensory thalamus in the processing of sensory information. In: Holstege G (ed) Progress in brain research. Elsevier, New York, pp 13–28
Ralston HJ, Ralston DD (1992) The primate dorsal spinothalamic tract: evidence for a specific termination in the posterior nuclei(Po/SG) of the thalamus. Pain 48:107–118
Ralston HJ, Ralston DD (1993) Local circuit processing in the primate thalamus: neurotransmitter mechanisms. In: Minciacchi D, Molinari M, Macchi G, Jones EG (eds) Thalamic networks for relay and modulation. Pergamon, Oxford, pp 109–122
Ralston HJ, Ralston DD (1994a) Medial lemniscal and spinal projections to the macaque thalamus: an electron microscopic study of differing GABAergic circuitry serving thalamic somatosensory mechanisms. J Neurosci 14:2485–2502
Ralston HJ, Ralston DD (1994b) GABA modulation of afferent information in the macaque somatosensory thalamus. FASEB J 8:A889
Reh T, Kalil K (1981) Development of the pyramidal tract in the hamster. I. A light microscopic study. J Comp Neurol 200:55–67
Remy P, Zilbovicius M, Leroywillig A, Syrota A, Samson Y (1994) Movement related and task related activations of motor cortical areas: a positron emission tomographic study. Ann Neurol 36:19–26
Richard AF (1985) Primates in nature. Freeman, New York
Robinson AH, Sale RD, Morrison JL, Muehrcke PC (1984) Elements of cartography. Wiley, New York
Robinson DL (1993) Functional contributions of the primate pulvinar. Prog Brain Res 95:371–380
Robinson DL, Goldberg ME (1978) Sensory and behavioural properties of neurons in posterior parietal cortex of the awake, trained monkey. Fed Proc 37:2258–2261
Robinson DL, Peterson SE (1992) The pulvinar and visual salience. TINS 15:127–132
Rockland K (1994) Further evidence for two types of corticopulvinar neurons. NeuroReport 5:1865–1868

Rodieck RW, Watanabe M (1993) Survey of the morphology of macaque retinal ganglion cells that project to the pretectum, superior colliculus, and parvicellular laminae of the lateral geniculate nucleus. J Comp Neurol 338:289–303

Roland PE, Larsen B, Lassen NA, Skinhoj E (1980a) Supplementary and other cortical areas in organization of voluntary movements in man. J Neurophysiol 43:118–136

Roland PE, Skinhoj E, Lassen NA, Larsen B (1980b) Different cortical areas in man in organization of voluntary movements in extrapersonal space. J Neurophysiol 43:137–150

Rose JE, Woolsey CN (1949) Organization of the mammalian thalamus and its relationships to the cerebral cortex. EEG Clin Neurophysiol 1:391–404

Rouiller EM, Moret V, Liang FY (1993) Comparison of the connectional properties of the 2 forelimb areas of the rat sensorimotor cortex: support for the presence of a premotor or supplementary motor cortical area. Somatosens Mot Res 10:269–289

Rouiller EM, Liang F, Babalian A, Moret V, Wiesendanger M (1994) Cerebellothalamocortical and pallidothalamocortical projections to the primary and supplementary motor cortical areas a multiple tracing study in macaque monkeys. J Comp Neurol 345:185–213

Ruch TC, Fulton JF, German WG (1938) Sensory discrimination in monkey, chimpanzee and man after lesions of the parietal lobe. Arch Neurol Psychiat 39:919–938

Rumelhart DE, Hinton GE, McClelland JL (1986) A general framework for parallel distributed processing. In: Rumelhart DE, McClelland JL (eds) MIT Press, Cambridge, MA, pp 45–76

Russell JR, DeMeyer W (1961) The quantitative cortical origin of pyramidal axons of Macaca rhesus. Neurology 11:96–108

Sachs E (1909) On the structure and functional relations of the optic thalamus. Brain 32:95–136

Sakai ST, Inase M, Tanji J (1994) Comparison of cerebellothalamic and pallidothalamic projections in the monkey. J Neurosci 100:100 (abstr)

Schmahmann JD, Pandya DN (1990) Anatomical investigation of projections from thalamus to posterior parietal cortex in the rhesus monkey: a WGA-HRP and fluorescent tracer study. J Comp Neurol 295:299–326

Schell GR, Strick PL (1984) The origin of thalamic inputs to the arcuate premotor and supplementary motor areas. J Neurosci 4:539–560

Schreyer DJ, Jones EG (1982) Growth and target finding by axons of the corticospinal tract in prenatal and postnatal rats. Neuroscience 7:1837–1853

Schreyer DJ, Jones EG (1988) Topographic sequence of outgrowth of corticospinal axons in the rat: a study using retrograde axonal labeling with fast blue. Dev Brain Res 466:89–101

Schwartz AB, Kettner RE, Georgopoulos AP (1988) Primate motor cortex and free arm movements to visual targets in three dimensional space. I. Relations between single cell discharge and direction of movement. J Neurosci 8(8):2913–2927

Schwartz ML, Goldman-Rakic PS (1991) Prenatal specification of callosal connections in rhesus monkey. J Comp Neurol 307:144–162

Seitz RJ, Roland PE (1992) Learning of sequential finger movements in man a combined kinematic and positron emission tomography (PET) study. Eur J Neurosci 4:154–165

Seitz RJ, Roland PE, Bohm C, Greitz T, Stoneelander S (1991) Somatosensory discrimination of shape tactile exploration and cerebral activation. Eur J Neurosci 3:481–492

Seitz RJ, Knorr U, Huang Y, Weder B, Herzog H, Schlaug G (1993) Comparison of intersubject averaging and individual response identification for mapping of motor function in the human brain with positron emission tomography (PET). Int Congr Ser 1030:561–568

Selemon L, Goldman-Rakic PS (1988) Common cortical and subcortical targets of the dorsolateral prefrontal and posterior parietal cortices in the rhesus monkey: evidence for a distributed neural network subserving spatially guided behaviour. J Neurosci 8:4049–4068

Seltzer B, Pandya DN (1989) Intrinsic connections and architectonics of the superior temporal sulcus in the rhesus monkey. J Comp Neurol 290:451–471

Seltzer B, Pandya DN (1994) Parietal, temporal and occipital projections to cortex of the superior temporal sulcus in the thesus monkey: a retrograde tracer study. J Comp Neurol 343:445–463

Semmes J, Weinstein S, Ghent L, Teuber H (1963) Correlates of impaired orientation in personal and extrapersonal space. Brain 86:747–772

Sessle BJ, Weisedanger M (1982) Structural and functional definition of the motor cortex in the monkey (Macaca fascicularis). J Physiol (Lond) 323:245–265

Sherrington C (1947) The integrative action of the nervous system. Cambridge University Press, Cambridge

Sherrington CS (1898) Decerebrate rigidity and reflex co-ordination of movements. J Physiol 22:319–332

Shibasaki H, Sadato N, Lyshkow H, Yonekura Y, Honda M, Nagamine T, Suwazono S, Magata Y, Ikeda A, Miyazaki M, Fukuyama H, Asato R, Konishi J (1993) Both primary motor cortex and supplementary motor area play an important role in complex finger movement. Brain 116:1387–1398

Shima K, Aya K, Mushiake H, Inase M, Aizawa H, Tanji J (1991) Two movement-related foci in the primate cingulate cortex observed in signal-triggered and self-paced forelimb movements. J Neurophysiol 65:188–202

Shinoda Y, Zarzecki P, Asanuma H (1979) Spinal branching of pyramidal tract neurons in the monkey. Exp Brain Res 34:59–72

Shinoda Y, Yokota J, Futami T (1981) Divergent projection of individual corticospinal axons to motoneurons of multiple muscles in the monkey. Neurosci Lett 23:7–12

Shinoda Y, Futami T, Mitoma H, Yokata J (1988) Morphology of single neurones in the cerebello-rubrospinal system. Behav Brain Res 28:59–64

Spreafico R, Frassoni C, Regondi MC, Arcelli P, De Biasi S (1993) Interneurons in the mammalian thalamus: a marker of species? In: Minciacchi D, Molinari M, Macchi G, Jones EG (eds) Thalamic networks for relay and modulation. Pergamon, Oxford, pp 17–28

Stanfield BB (1992) The development of the corticospinal projection. Prog Neurobiol 38:169–202

Stanfield BB, O'Leary DDM (1985) Fetal occipital cortical neurons transplanted to the rostral cortex can extend and maintain a pyramidal tract axon. Nature 313:135–137

Stanfield BB, O'Leary DDM, Fricks C (1982) Selective collateral elimination in early postnatal development restricts cortical distribution of rat pyramidal tract neurones. Nature 298:371–373

Stein JF (1992) The representation of egocentric space in the posterior parietal cortex. Behav Brain Sci 15:691–700

Stelmach GE, Castiello U, Jeannerod M (1994) Orienting the finger opposition space during prehension movements. J Motor Behav 26:178–186

Steriade M, Pare D, Parent A, Smith Y (1988) Projections of cholinergic and noncholinergic neurons of the brainstem core to relay and associational thalamic nuclei in the cat and macque monkey. Neuroscience 25:47–67

Steriade M, Contreras D, Curo Dossi R, Nunez A (1993a) The slow (<1 Hz) oscillation in reticular thalamic and thalamocortical neurons: scenario of sleep rhythm generation in interacting thalamic and neocortical networks. J Neurosci 13:3284–3299

Steriade M, McCormick DA, Sejnowski TJ (1993b) Thalamocortical oscillations in the sleeping and aroused brain. Science 262:679–685

Strick P (1975) Multiple sources of thalamic input to the primate motor cortex. Brain Res 88:372–377

Strick P (1976) Anatomical analysis of ventrobasal thalamic input to the primate motor cortex. J Neurophysiol 39:1020–1031

Talbot WH, Darian-Smith I, Kornhuber HH, Mountcastle VB (1968) The sense of flutter-vibration: comparison of the human capacity with response patterns of mechanoreceptive afferents from the monkey hand. J Neurophysiol 31:301–334

Tanji J (1994) The supplementary motor area in the cerebral cortex. Neurosci Res 19:251–268

Tanji J, Kurata K (1983) Functional organization of the supplementary motor area. In: Desmedt JE (ed) Motor control mechanisms in health and disease. Raven, New York, pp 421–431

Tanji J, Kurata K (1985) Contrasting neuronal activity in supplementary and precentral motor cortex of monkeys. I. Responses to instructions determining motor responses to forthcoming signals of different modalities. J Neurophysiol 53:129–141

ten Donkelaar HJ (1988) Evolution of the red nucleus and rubrospinal tract. Behav Brain Res 28:9–20

Thach WT (1978) Correlation of neural discharge with pattern and force of muscular activity, joint position, and direction of intended next movement in motor cortex and cerebellum. J Neurophysiol 41:654–676

Thach WT, Goodkin HP, Keating JG (1992) The cerebellum and the adaptive coordination of movement. Annu Rev Neursosci 15:403–442

Thach WT, Kane SA, Mink JW, Goodkin HP (1993) Cerebellar output: multiple maps and modes of control in movement coordination. In: Llinas R, Sotelo C (eds) The cerebellum revisited. Springer, Berlin Heidelberg New York, pp 283–300

Thielert CD, Thier P (1993) Patterns of projections from the pontine nuclei and the nucleus reticularis tegmenti pontis to the posterior vermis in the rhesus monkey a study using retrograde tracers. J Comp Neurol 337:113–126

Tigges J, Nakagawa S, Tigges M (1979) Efferents of area 4 in a south american monkey (Saimiri). 1. Terminations in the spinal cord. Brain Res 171:1–10

Tokuno H, Tanji J (1993) Input organization of distal and proximal forelimb areas in the monkey primary motor cortex: a retrograde double labeling study. J Comp Neurol 333:199–209

Tokuno H, Kimura M, Tanji J (1992) Pallidal inputs to thalamocortical neurons projecting to the supplementary motor area: an anterograde and retrograde double labeling study in the macaque monkey. Exp Brain Res 90:635–638

Toyoshima K, Sakai H (1982) Exact cortical extent of the origin of the corticospinal tract (CST) and the quantitative contribution to the CST in different cytoarchitectonic areas. A study with horseradish peroxidase in the monkey. J Hirnforsch 23:257–269

Tracey DJ, Asanuma C, Jones EG, Porter R (1980) Thalamic relay to motor cortex: afferent pathways from brain stem, cerebellum, and spinal cord in monkeys. J Neurophysiol 44:532–554

Turman AB, Ferrington DG, Ghosh S, Morley JW, Rowe MJ (1992) Parallel processing of tactile information in the cerebral cortex of the cat: effect of reversible inactivation of SI on responsiveness of SII neurons. J Neurophysiol 67:411–429

Ungerleider W, Mishkin M (1982) Two cortical visual systems. In: Ingel DJ, Mansfield RJW, Goodale MA (eds) The analysis of visual behavior. MIT Press, Cambridge, pp 549–586

Usrey WM, Fitzpatrick D (1993) Parallel streams in the corticogeniculate pathway: intrinsic and extrinsic projections of neurons in layer VI of striate cortex. In: Minciacchi D, Molinari M, Macchi G, Jones EG (eds) Thalamic networks for relay and modulation. Pergamon, Oxford, pp 91–98

Van Gelder T (1984) Playing flourens to fodor gall. Behav Brain Sci 17: 84

Vanderlinden C, Bruggeman R (1993) Multiple descending corticospinal volleys demonstrated by changes of the wrist flexor H reflex to magnetic motor cortex stimulation in intact human subjects. Muscle Nerve 16:374–378

Velayos JL, Casas-Puig R, Reinoso-Suarez F (1993) Laminar organization of the cortical projections to the intralaminar and medial thalamic nuclei in the cat. In: Minciacchi D, Molinari M, Macchi G, Jones EG (eds) Thalamic networks for relay and modulation. Pergamon, Oxford, pp 185–196

Vogt BA, Pandya DN (1987) Cingulate cortex of the rhesus monkey. II. Cortical afferents. J Comp Neurol 262:271–289

Vogt BA, Pandya DN, Rosene DL (1987) Cingulate cortex of the rhesus monkey. I. Cytoarchitecture and thalamic afferents. J Comp Neurol 262:256–270

Vogt C (1909) La myeloarchitecture du thalamus du cercopithique. J Psychol Neurol 12: 285–324

Vogt C, Vogt O (1919) Allgemeinere Ergebnisse unserer Hirnforschung. J Psychol Neurol 25:279

von Bonin G, Bailey P (1947) The neocortex of the Macaca mulatta. University of Illinois Press, Urbana

von Kolliker A (1896) Handbuch der Gewebelehre des Menschen. Engelmann, Leipzig

Walker AE (1938) The primate thalamus. University of Chicago Press, Chicago

Walker AE (1939) The origin, course and terminations of the secondary pathways of the secondary pathways of the trigeminal nerve in primates. J Comp Neurol 71:59–89

Wassermann EM, Pascualleone A, Hallett M (1994) Cortical motor representation of the ipsilateral hand and arm. Exp Brain Res 100:121–132

Watson JDG, Myers R, Frackowiak RSJ, Hajnal JV, Woods RP, Mazziotta JC, Shipp S, Zeki S (1993) Area V5 of the human brain evidence from a combined study using positron emission tomography and magnetic resonance imaging. Cerebr cortex 3:79–94

Webster MJ, Bachevalier J, Ungerleider LG (1994) Connections of inferior temporal areas TEO and TE with frontal and parietal cortex in macaque monkeys. Cerebr Cortex 4: 470–483

Weder B, Knorr U, Herzog H, Nebeling B, Kleinschmidt A, Huang Y, Steinmetz H, Freund JH, Seitz RJ (1994) Tactile exploration of shape after subcortical ischemic infarction studied with PET. Brain 117:593–605

Weiller C, Chollet F, Friston KJ, Wise RJS, Frackowiak RSJ (1992a) Functional reorganization of the brain in recovery from striatocapsular infarction in man. Ann Neurol 31:463–472

Weiller C, Ramsay SC, Friston KJ, Frackowiak RSJ (1992b) Reshaping the human cortical motor map after stroke. Ann Neurol 32:232

Wernicke C (1881) Lehrbuch der Gehirnkrankheiten. Kassel

Wessel K, Zeffiro T, Toro C, Hallett M (1994) Activation of different brain areas during self paced or metronome paced finger movements a PET study. Neurology 44:A352

Wiesendanger R, Wiesendanger M (1985) The thalamic connections with medial area 6 (supplementary motor cortex) in the monkey (Macaca fascicularis). Exp Brain Res 59: 91–104

Wiesendanger M, Hummelsheim H, Bianchetti M, Chen DF, Hyland B, Maier V, Wiesendanger R (1987) Input and output organization of the supplementary motor area. Ciba Found Symp 132:40–53

Williamson JR, Ralston HJI (1993) There are few dendro-dendritic synaptic contacts in primate thalamic reticular nucleus (TRN). Soc Neurosci Abstr 19(1):516

Willis WD, Coggeshall RE (1991) Sensory mechanisms of the spinal cord. Plenum, New York

Windhorst UR (1991) What are the output units of motor behaviour and how are they controlled? In: Humphrey DR, Freund H (eds) Motor control: concepts and issues. Dahlem workshop report. Wiley, Chichester, pp 101–119

Woolsey CN, Settlage PH, Meyer DR, Spencer W, Hamuy TP, Travis AM (1952) Patterns of localization in precentral and "supplementary" motor areas and their relation to the concept of a premotor area. Res Publ Assoc Nerv Ment Dis 30:238–264

Yen C-T, Jones EG (1983) Intracellular staining of physiologically identified neurons and axons in the somatosensory thalamus of the cat. Brain Res 280:148–154

Yen C-T, Conley M, Jones EG (1985a) Morphological and functional types of neurons in cat ventral posterior thalamic nucleus. J Neurosci 5:1316–1338

Yen C-T, Hendry SHC, Jones EG (1985b) The morphology of physiologically identified GABAergic neurons in the somatic sensory part of the thalamic reticular nucleus in the cat. J Neurosci 5:2254

Yeterian EH, Pandya DN (1991) Corticothalamic connections of the superior temporal sulcus in rhesus monkeys. Exp Brain Res 83:268–284

Yeterian EH, Pandya DN (1994) Laminar origin of striatal and thalamic projections of the prefrontal cortex in rhesus monkeys. Exp Brain Res 99:383–398

Zarzecki P (1991) The distribution of corticocortical, thalamocortical, and callosal inputs on identified motor cortex output neurons mechanisms for their selective recruitment. Somatosens Mot Res 8:313–325

Zeki S, Shipp S (1988) The functional logic of cortical connections. Nature 335:311–317

Zeki S, Watson JDG, Lueck CJ, Friston KJ, Kennard C, Frackowiak RSJ (1991) A direct demonstration of functional specialization in human visual cortex. J Neurosci 11:641–649

Zola-Morgan S (1995) Localization of brain function: the legacy of Franz Joseph Gall (1758–1828). Annu Rev Neurosci 18:359–384

7 Subject Index

Springer-Verlag and the Environment

We at Springer-Verlag firmly believe that an international science publisher has a special obligation to the environment, and our corporate policies consistently reflect this conviction.

We also expect our business partners – paper mills, printers, packaging manufacturers, etc. – to commit themselves to using environmentally friendly materials and production processes.

The paper in this book is made from low- or no-chlorine pulp and is acid free, in conformance with international standards for paper permanency.